INTEGRATED AUDIO AMPLIFIERS IN BCD TECHNOLOGY

INTEGRATED AUDIO AMPLIFIERS IN BCD TECHNOLOGY

by

Marco Berkhout
*MESA Research Institute,
University of Twente,
and
Philips Semiconductors*

Kluwer Academic Publishers
BOSTON / DORDRECHT / LONDON

A C.I.P. Catalogue record for this book is available from the Library of Congress

ISBN 0-7923-8003-7

Published by Kluwer Academic Publishers,
P.O. Box 17, 3300 AA Dordrecht, The Netherlands.

Sold and distributed in the U.S.A. and Canada
by Kluwer Academic Publishers,
101 Philip Drive, Norwell, MA 02061, U.S.A.

In all other countries, sold and distributed
by Kluwer Academic Publishers,
P.O. Box 322, 3300 AH Dordrecht, The Netherlands.

Printed on acid-free paper

All Rights Reserved
© 1997 Kluwer Academic Publishers
No part of the material protected by this copyright notice may be reproduced or
utilized in any form or by any means, electronic or mechanical,
including photocopying, recording or by any information storage and
retrieval system, without written permission from the copyright owner.

Printed in the Netherlands

"Sounds like art," the instructor says.
"Well, it is art," I say. "This divorce of art from technology is completely unnatural. It's just that it's gone on so long you have to be an archeologist to find out where the two separated."

>Robert M. Pirsig
>
>*Zen and the Art of Motorcycle Maintenance*

Contents

Foreword	xi
Preface	xiii
1 Introduction	**1**

1.1 Integrated Audio Amplifiers	1
1.1.1 Output Power and Dissipation	2
1.1.2 Amplifier Classes	4
1.1.3 Amplifier Requirements	10
1.2 BCD Technology	11
1.2.1 DMOS versus Bipolar	12
1.2.2 DMOS Output Stage	13
1.3 Outline of this Thesis	14
1.4 References	15

2 DMOS Technology	**17**

2.1 Introduction	17
2.1.1 Historical Perspective	17
2.1.2 Introduction to DMOS	18
2.1.3 Application Areas	22
2.2 DMOS Device Structures	23
2.2.1 Lateral DMOS	23
2.2.2 Vertical DMOS	27
2.2.3 VMOS and UMOS	32

2.2.4 Comparison of DMOS Structures — 35
2.3 DMOS Device Properties — 36
   2.3.1 Transfer Characteristics — 36
   2.3.2 Parasitic Elements — 38
2.4 DMOS Process Technology — 41
   2.4.1 Isolation Techniques — 41
   2.4.2 Device Termination Techniques — 44
   2.4.3 BCD Technology — 45
2.5 DMOS Versus Bipolar — 52
   2.5.1 Second Breakdown — 52
   2.5.2 Bipolar/DMOS Hybrids — 55
2.6 Conclusion — 57
2.7 References — 57

3 Chargepump Circuits — 63

3.1 Introduction — 63
   3.1.1 Switching Voltage Regulators — 64
   3.1.2 Voltage Multiplication — 64
   3.1.3 Application Areas of Chargepumps — 66
3.2 Voltage Multipliers — 67
   3.2.1 Marx Voltage Multiplier — 68
   3.2.2 Cockcroft-Walton Voltage Multiplier — 70
   3.2.3 Dickson Voltage Multiplier — 72
   3.2.4 Comparison of Voltage Multipliers — 73
3.3 Chargepump Operation — 73
   3.3.1 Normal Mode — 75
   3.3.2 Double Phase Mode — 75
   3.3.3 Current Driven Mode — 77
   3.3.4 Double Phase Current Driven Mode — 78
   3.3.5 Simulation Results — 80

3.4 Voltage Control	80
3.4.1 Output Voltage Clipping	81
3.4.2 Amplitude Control	81
3.4.3 Frequency Control	82
3.4.4 Active Stage Control	82
3.5 Chargepump Implementation	82
3.5.1 Diodes and Switches	82
3.5.2 Driver Circuits	84
3.5.3 Output Voltage Detection	85
3.5.4 Realization Example	86
3.6 Conclusion	88
3.7 References	89
3.8 Appendix: Down Conversion	91
3.8.1 Voltage Division	91
3.8.2 References	96

4 Chargepump Modeling   97

4.1 Introduction	97
4.1.1 Chargepump Models	98
4.2 Extended Chargepump Model	100
4.2.1 Some Notation Conventions	101
4.2.2 Transient Modeling	103
4.2.3 Steady-State Modeling	107
4.2.4 Transient Behavior	109
4.3 Modeling of Parasitic Effects	111
4.3.1 Current Leakage	112
4.3.2 Series Resistance	114
4.3.3 Stray Capacitance	119
4.3.4 Parallel Capacitance	122
4.3.5 Body-Effect	127

4.3.6 Combination of Parasitic Effects	131
4.4 Conversion Efficiency	132
4.4.1 Power Consumption and Efficiency	132
4.4.2 Influence of Parasitics	134
4.5 Conclusion	136
4.6 References	136

5 BCD Audio Amplifiers 137

5.1 Introduction	137
5.1.1 Sources of Distortion	138
5.1.2 Output Resistance	142
5.1.3 Circuit Design in BCD Technology	145
5.2 Output Stage Topologies	146
5.2.1 Common Drain Stages	147
5.2.2 Common Source Stages	154
5.2.3 Comparison of Output Stages	161
5.3 A BCD Amplifier Design	163
5.3.1 Amplifier Topology	163
5.3.2 Signal Splitter	169
5.3.3 Input Stage	184
5.3.4 Output Drivers	185
5.3.5 Complete Amplifier	196
5.4 Conclusion	200
5.5 References	201

6 Conclusion 203

6.1 Conclusions	203
6.2 Recommendations	206

Index

Foreword

Audio power amplifiers were among the first analog circuits available as integrated circuits. Early on, there was a lot of similarity with operational amplifiers, but during its evolution, the audio power amplifier increasingly became a special art of analog electronics. The modern integrated audio amplifier can deliver an audio output power up to 100 Watt and is robust against almost any kind of mishandling such as a short circuit across the load or to the supply lines. Besides this, it has a very high ratio between quiescent current and maximum output current, very good linearity and the ability to handle almost any complex load without oscillation. Audio amplifiers are often the last link in the audio chain, directly connected to the loudspeaker, and therefore common mode input signals, supply voltage variations and switch-on should not give any audible output from the loudspeaker.

Until recently almost all integrated audio amplifiers were designed in bipolar processes. Although it is possible to design very good bipolar amplifiers, there are some drawbacks that limit the performance for future generations of audio amplifiers. The first limitation is in the bipolar power transistor. For high voltage and high power the Safe Operating ARea (SOAR) of the bipolar transistor is a serious limitation to the design of a robust integrated high power audio amplifier. Besides the bipolar process is not well suited to the integration of digital and mixed signal circuits such as buses for control and diagnoses and digital to analog converters. In a world where digital audio signals have become standard, integrating these functions with the audio amplifier is a logical choice, especially where the submicron CMOS processes for the DSP elements is not particularly suitable for high performance analog audio signals. To overcome the limitations of the bipolar process a new process technology needs to be used. Marco Berkhout's book concentrates on the design of the analog part of a integrated audio power amplifier in a Bipolar CMOS DMOS (BCD) process. The DMOS power transistor, with its very good SOAR, offers a robust, compact alternative for the bipolar power transistor. Besides the CMOS opens the way to the addition of digital and mixed-signal circuits.

The DMOS power transistor has a large and small signal behavior that differs considerably from that of the bipolar power transistor. Furthermore,

Foreword

in the small signal part of the amplifier the bipolar transistor is non preferred because its quality becomes inferior in newer BCD processes. Another particularity is that the upper output transistor needs a boosted supply voltage to realize a rail to rail output voltage. For maximum output power the DMOS power transistor is used in its saturated and in its non-saturated region resulting in an extra source of non linearity that is unknown in bipolar amplifiers. All these differences between bipolar and BCD amplifiers gave rise to the challenge to design new basic circuits that satisfy the high standards of audio power amplifiers. A number of the basic circuits get special attention in this book.

The book starts with an extensive discussion of the properties of the DMOS transistor. Then the theory and the design of the charge pump that is needed for the boosted supply voltage is considered. The new solutions that are found can also be used for many applications where DC-DC conversion with low output ripple is needed. The design of the amplifier concentrates on a new quiescent control circuit with very high ratio between quiescent current and maximum output current and on the output stage topologies. The problem of controlling the DMOS output transistors over a wide range of currents either saturated or non saturated requires a completely new design of the driving circuits that utilize of the special properties of the DMOS transistor. All of this is explained in a clear way and it will certainly help the reader to set foot down the difficult path of audio amplifier design in BCD technology.

Ed van Tuijl

Preface

In this book the design of a fully integrated *100W* audio power amplifier in a *BCD* technology is presented. In a *BCD* technology a combination of Bipolar, *CMOS* and *DMOS* transistors is available.

An attempt is made to develop a design strategy that can be applied to *BCD* technologies in general. However, since the designs presented in this book are all realized in one specific *BCD* technology, many design choices are based on the characteristics of this technology.

Key elements in the amplifier design are a fully integrated chargepump circuit and a common source output stage. The overall design goal is to achieve high open loop linearity.

A number of amplifier classes is presented and compared with respect to power dissipation. A list of requirements is given that an integrated audio amplifier has to satisfy. Further, the necessity of a voltage higher than the supply voltage in order to achieve rail-to-rail output capability is explained.

The various aspects of *DMOS* are explored. After a short introduction to the role of *DMOS* in integrated circuit technology today, a qualitative overview is presented of the device physics of *DMOS*, the best known *DMOS* device structures and the technological aspects of *DMOS*. A comparison is made between *DMOS* and bipolar transistors.

Voltage multipliers are circuits that can generate a voltage higher than the supply voltage without the use of inductors. Some well-known voltage multiplication techniques are presented and compared on their suitability for integration. Based on this comparison a specific voltage multiplier circuit called a *chargepump* is selected. A technique is presented to reduce the output voltage ripple of chargepumps. Some methods are discussed for regulation of the output voltage level. A fully integrated chargepump design is presented and discussed.

A detailed model that describes the operation of chargepump circuits is presented. First, a brief overview of previously published chargepump models is given. Next, the development of a new model is described that is based upon analysis of the charge balance between adjacent capacitors in the chargepump. With this model both transient and steady-state behavior

Preface

of chargepumps can be described accurately. It is demonstrated that the influence of a number of parasitics can easily by included in the model.

An integrated *100W* audio power amplifier is presented realized in a *BCD* technology. Several different amplifier topologies are discussed and compared on their suitability for integration in a *BCD* technology. Based on this comparison one particular amplifier topology is selected and developed further. A detailed description of the design of this amplifier is presented in which a chargepump circuit is used in order to achieve rail-to-rail output capability.

The dominant source of distortion turns out to be the large input capacitance of the output transistors. The largest part of the input capacitance is formed by the gate-drain capacitance. This capacitance increases substantially when the drain and gate voltage decreases due to a accumulation phenomenon that is particular to Vertical *DMOS* transistors. Further, due to the Miller-effect the gate-drain capacitance appears to be even larger.

The work presented in this book is the result of a four-year research period that was performed at the MESA Research Institute, University of Twente in close collaboration with Philips Semiconductors Nijmegen.

Introduction

Monolithic audio power amplifiers are widely used in car-stereo, television sets, portable audio systems and basically any application where a few tenths to tens of watts are needed in the *20Hz* to *20kHz* frequency range [1]. Recent developments in the consumer electronics field show a tendency towards higher output powers (*~100W*) and hi-fi performance comparable to discrete designs. In many popular musical styles, the physical impact of the sound is an important part to the enjoyment of the music. In order to double the subjective loudness, the output power of an amplifier should be increased tenfold [2]. On the other hand, the advent of digital signal sources such as the Compact Disc has put severe requirements on distortion specifications. Consequently, design of audio power amplifiers revolves around high output power and low distortion.

1.1 Integrated Audio Amplifiers

High efficiency is a desirable feature for audio power amplifiers. Clearly, high efficiency is desirable from an economic point of view. This is especially true in battery powered applications such as portable audio sets. Higher efficiency increases the lifespan of the batteries. However a more important reason for pursuing high efficiency is reduction of the dissipation in the amplifier output stage. Heat generation caused by

Introduction

dissipation can deteriorate the performance and can lead to destruction of the amplifier. Therefore, dissipation can be a limiting factor in power output capability.

1.1.1 Output Power and Dissipation

Most integrated power amplifiers have two output transistors in single-ended *(SE)* push-pull configuration and use a unipolar supply voltage V_P as indicated in Figure 1.1.1. These transistors can be bipolar or *MOS* transistors of either polarity represented here by controlled current sources. One transistor is placed between the power supply V_P and output node and is called the *high-side* transistor. The other transistor is placed between the output node and ground and is called the *low-side* transistor. The load Z_L is connected between the amplifier output and a node at half the supply voltage $V_P/2$. In many applications a large external capacitor C_{ext} is used to realize this half supply voltage $V_P/2$ for AC signals.

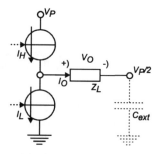

Figure 1.1.1 Single-ended (SE) Push-pull output configuration

Both transistors work together in so-called *push-pull* operation. The name push-pull is derived from the observation that signal current flowing out of the amplifier is "pushed" into the load by the high-side transistor whereas current flowing into the amplifier is "pulled" out of the load by the low-side transistor. The output current $I_O(t)$ is the difference between the current through the high-side transistor $I_H(t)$ and low-side transistor $I_L(t)$:

$$I_O(t) = I_H(t) - I_L(t)$$

(1.1.1)

The current that flows through the output transistors when the output current I_O is zero is called the *quiescent current* I_q.

The *instantaneous* output power $P_O(t)$ and dissipation $P_D(t)$ in a push-pull output stage are given by:

$$P_O(t) = I_O(t) \cdot V_O(t)$$
$$P_D(t) = I_H(t) \cdot \left(\frac{1}{2} \cdot V_P - V_O(t)\right) + I_L(t) \cdot \left(\frac{1}{2} \cdot V_P + V_O(t)\right)$$

(1.1.2)

The *average* values of the output power and dissipation depend on the output signal $V_O(t)$ and the load impedance Z_L. Usually, a sinewave is assumed in combination with an ohmic load R_L which leads to:

$$V_O(t) = V_O \cdot \sin(\omega \cdot t)$$
$$I_O(t) = \frac{V_O}{R_L} \cdot \sin(\omega \cdot t)$$

(1.1.3)

The *average* power dissipation P_D can be calculated with:

$$P_D = \frac{\omega}{2\pi} \int_0^{\frac{2\pi}{\omega}} P_D(t) \cdot dt$$

(1.1.4)

whereas the *average* output power P_O is simply given by:

$$P_O = \frac{V_O^2}{2 \cdot R_L}$$

(1.1.5)

The average output power of a full swing sinewave into a 4Ω or 8Ω load resistor is used to rate the maximum output power of an amplifier, e.g. *100W/8Ω*. For the output stage shown in Figure 1.1.1 the maximum signal amplitude is equal to $V_P/2$ so the maximum output power is:

$$P_{O,max} = \frac{V_P^2}{8 \cdot R_L}$$

(1.1.6)

The maximum output power of an amplifier can be quadrupled by using a bridge-tied-load (*BTL*) configuration as shown in Figure 1.1.2. In this configuration, the load is connected between the outputs of two identical amplifiers which are driven in antiphase. Consequently, the maximum output signal amplitude is doubled to V_P while at the same time the half supply voltage $V_P/2$ is not needed anymore. The dissipation in the output transistors is also quadrupled so the efficiency of a *BTL* amplifier is equal to that of an *SE* amplifier.

Introduction

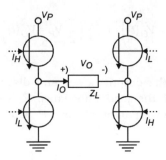

Figure 1.1.2 Output stage with Bridge-tied-load (BTL) configuration

The average dissipation can also be calculated if reactive loads are being used. A reactive load Z_L causes the load current I_O to lead or lag the load voltage V_O by phase ϕ_L:

$$I_O(t) = \frac{V_O}{|Z_L|} \cdot \sin(\omega \cdot t + \phi_L)$$

(1.1.7)

The average dissipation in the amplifier output stage is considerably higher with reactive loads than with resistive loads of the same magnitude [3].

An alternative method to calculate the average power dissipation is to use the amplitude distribution of the output signal. Real music signals do not have the same amplitude distribution as sinewaves but can often be represented with a Gaussian distribution [4]. However, because the amplitude distribution does not contain frequency information this method cannot be used if the load impedance is frequency dependent which usually is the case with real loudspeakers.

1.1.2 Amplifier Classes

In order to minimize dissipation in the output stage the current through and the voltage across the output transistors should be as low as possible at all time. The dependence of $I_H(t)$ and $I_L(t)$ on the output current is determined by the amplifier class. A brief summary of some amplifier classes is given in the following paragraphs.

Class A

Amplifiers with class A operation give the lowest distortion figures but also have the lowest efficiency. In class A both output transistors conduct current at all time. Each transistor supplies half of the signal current I_O. Consequently, the sum of the currents through the output transistors stays nearly constant. The quiescent current I_q is usually set to be half the maximal output current. The current through the output transistors in class A is given by:

$$I_H(t) = I_q + \frac{1}{2} \cdot I_O(t)$$
$$I_L(t) = I_q - \frac{1}{2} \cdot I_O(t)$$

(1.1.8)

The average dissipation P_D as a function of *virtual output power* P_V for a number of different load phases ϕ_L is shown in Figure 1.1.3. Here *virtual output power* P_V is defined as:

$$P_V = \frac{V_O^2}{2 \cdot |Z_L|}$$

(1.1.9)

The power levels are normalized by dividing by the maximum virtual output power $P_{V,max}$.

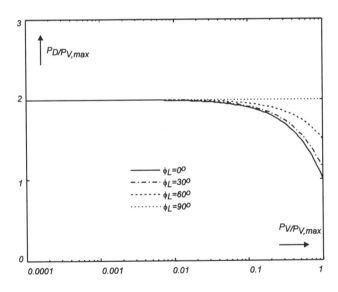

Figure 1.1.3 Average dissipation as a function of virtual output power in class A

As can be seen in Figure 1.1.3 the dissipation in a class A amplifier decreases when the output power increases. The minimum dissipation is reached at maximum output power with an ohmic load. In this case the dissipation equals the output power and efficiency is 50%. The curves are

Introduction

valid for both SE and BTL configurations. Clearly, class A is not a very attractive class for high power amplifiers.

Class B

A significant reduction in dissipation can be achieved with class B operation. In a class B amplifier either the high-side or the low-side transistor is conducting current while the other device is turned off. In pure class B there is no quiescent current so the quiescent dissipation is zero. Consequently, only one of the output transistors is dissipating power at any time.

$$\forall I_O(t) > 0 = \begin{cases} I_H(t) = I_O(t) \\ I_L(t) = 0 \end{cases}$$

$$\forall I_O(t) < 0 = \begin{cases} I_H(t) = 0 \\ I_L(t) = -I_O(t) \end{cases}$$

(1.1.10)

The average dissipation P_D as a function of virtual output power P_V for a number of different load phases ϕ_L is shown in Figure 1.1.4.

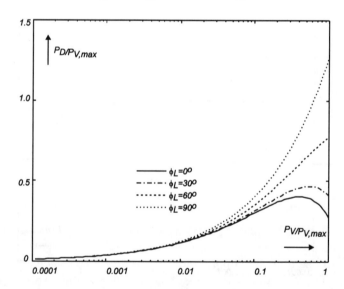

Figure 1.1.4 Average dissipation as a function of virtual output power in class B

As can be seen from Figure 1.1.4 the dissipation in a class B output stage is substantially lower than in class A. The class B operation is the cause of some typical forms of distortion. Best known is the so-called crossover distortion which occurs in the region where the active output transistor switches off while the other switches on. This generates a spray of higher order harmonics in the output signal. Another source of distortion is crosstalk of the nonlinear class B supply currents to nodes that carry linear signals. Finally, differences in gain between the high-side and low-side signal paths cause mainly even order harmonic distortion.

Reduction of crossover distortion can be achieved by using a more complicated biasing regime with a small quiescent current. The output transistors do not turn off completely but always conduct a small so-called *residual current*. In the crossover region both transistors contribute to the output current similar to class A operation. This biasing regime is called class AB. The dissipation is comparable to that in class B since the small quiescent and residual current only have a small contribution to the total dissipation.

Class G

A further reduction in dissipation can be achieved with class G operation. In a class G amplifier the voltage drop across the conducting output transistors is reduced by switching between multiple supply voltages.

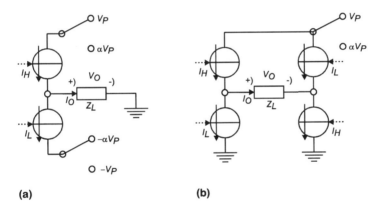

Figure 1.1.5 Class G Output stages (a) SE configuration (b) BTL configuration

Usually, class G amplifiers have a SE configuration and use multiple bipolar supply voltages as shown in Figure 1.1.5(a). However, it is also possible to use unipolar supply voltages by using a BTL configuration as is shown in Figure 1.1.5(b). Here two supply rails V_P and αV_P with $\alpha<1$ are used. The high-side transistors are only connected to the higher supply if this is required by the output signal. Consequently, the voltage across the output transistors is reduced part of the time which results in lower dissipation.

The value of α influences the decrease in dissipation that can be achieved. The optimal value can be approximated when the amplitude distribution of

Introduction

the output signal is known or some worst case distribution is assumed [4]. The average dissipation P_D for $\alpha=0.5$ as a function of virtual output power P_V for a number of different load phases ϕ_L is shown in Figure 1.1.6. Again a considerable reduction in dissipation can be observed compared to class B.

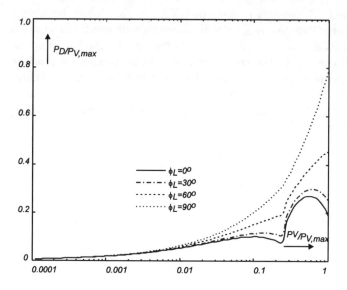

Figure 1.1.6 Average dissipation as a function of virtual output power in class G

The dissipation can be reduced further if more supply rails are used. However, this also requires a larger number of switches and increases the complexity of the amplifier.

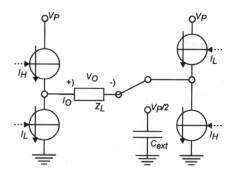

Figure 1.1.7 Switched bridge output stage

Integrated Audio Amplifiers

An alternative class G output stage that does not require a second supply voltage is shown in Figure 1.1.7. For small signals the amplifier operates in SE mode and the load is connected to the coupling capacitor C_{ext}. If the signal amplitude exceeds $V_P/2$ the switch connects the second amplifier stage and the amplifier operates in BTL mode.

The average dissipation curves for this output stage are identical to that shown in Figure 1.1.6. The advantage of this stage is that it does not require additional supply rails. However, the switch has to be able to conduct current in both directions and is directly in series with the load which can make it a significant source of distortion.

Class D

The highest possible efficiency is achieved with class D operation. In a class D output stage the output transistors are used as switches that connect the load alternately to the supply rail or to ground. Consequently, dissipation only occurs during the steep transients. The output signal of a class D amplifier is a pulse-width-modulated (PWM) signal that can be constructed by comparing the input signal with a triangular signal as shown in Figure 1.1.8.

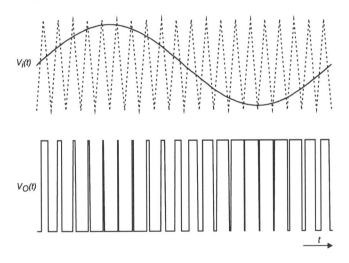

Figure 1.1.8 Construction of class D output signal

The loudspeaker signal can be extracted from the output signal with a low loss LC lowpass filter which cannot be integrated. Further, the high speed switching can cause interference and EM-radiation. For high fidelity performance the timing of the switches has to be controlled very accurately which is very difficult.

Introduction

Other Classes

In the previous overview the classes *C,E* and *F* are missing. Although these classes exhibit high efficiency they are not suited for audio due to their nonlinearity. Application of these classes can be found in high frequency applications such as radio transmitters and switching regulators.

Further, many hybrid classes are possible in which amplifiers of different types are combined. In general a coarse high efficiency amplifier is combined with an accurate low efficiency amplifier in order to make a compromise between efficiency, accuracy and complexity. However, these amplifiers are beyond the scope of this book.

1.1.3 Amplifier Requirements

In this section some requirements that are put on modern integrated audio power amplifiers are given. These requirements apply to a single-ended class *AB* amplifier. This is a reasonable restriction since many circuit parts of the *SE* class *AB* amplifier are also used in other amplifier configurations such as *BTL* and class *G*.

Output Power

The maximum output power of an amplifier is limited by the supply voltage. For a single-ended amplifier with a single supply V_P and ohmic load R_L this maximum output power is:

$$P_{O,max} = \frac{V_P^2}{8 \cdot R_L}$$

(1.1.11)

where it is assumed that the output of the amplifier can be driven from rail to rail. Usually, this is not the case because the output transistors require a certain voltage in order to operate. The output stage must be designed in such a way that the output window is as large as possible. For a supply voltage of *60V* an output power of *100W/4Ω* is possible.

Quiescent Current

The bias current of the amplifier must be as low as possible to minimize dissipation. A reasonable value for a *100W* amplifier is in the order of a few tens of milliAmperes. In this book a value of *25mA* is aimed for. About *10mA* is reserved for quiescent current through the output transistors while the rest can be divided between input stage and driver circuits.

Closed Loop Gain

The closed loop gain for power amplifiers is usually somewhere in the range of *26dB* to *30dB*. The closed loop gain is determined by the feedback resistors. In the design presented in this book the feedback network is not integrated with the amplifier. Therefore, the exact value for the closed loop gain can be chosen later.

Output Resistance

The output resistance must be low for good loudspeaker damping and preferably be constant over the audio band. In general, it is not useful to put much effort into reaching values lower than *100mΩ* since the resistance of loudspeaker cables is in the same order of magnitude.

Distortion

Distortion in audio amplifiers is usually determined by using a single sinewave as test signal and measuring the total harmonic distortion *(THD)* in the audio band. The measured *THD* depends on the both the amplitude and frequency of the sine wave and the load impedance. Usually a frequency in the range of *1kHz* to *5kHz* in combination with a *4Ω* to *8Ω* load is used and *THD* is measured as a function of output power. The *THD* of a *100W/4Ω* amplifier should be well below *0.1%* up to the onset of clipping at the output.

Crossover distortion is especially audible at low signal levels. Therefore, crossover distortion should be as low as possible.

Slew-rate

An audio power amplifier must be able to reproduce a full swing *20kHz* sine wave without excessive distortion. This corresponds to a slew-rate *SR* of:

$$SR = \frac{V_P}{2} \cdot 2\pi \cdot 20.000$$

(1.1.12)

where V_P is the supply voltage. For a supply voltage of *60V* this leads to a slew-rate of about *4V/μs*.

Stability

Ideally, an amplifier must be electrically stable and free from oscillations with all passive loads from purely inductive to purely capacitive. However, many real integrated amplifiers do not satisfy this requirement without the addition of a compensation network at the output.

1.2 BCD Technology

Most integrated power amplifiers today are realized in bipolar technologies using power *NPN* transistors in the output stage. This preference is caused mainly by the availability of these transistors in many technologies. This situation has changed due to the appearance of *BCD* technologies during

Introduction

the 1980's. In a *BCD* technology *B*ipolar *C*MOS and *D*MOS transistors can be integrated on the same chip. This development has made it feasible to use *DMOS* transistors as an alternative for bipolar transistors in fully integrated audio amplifiers.

1.2.1 DMOS versus Bipolar

Many things can be and have been said in favor or at the expense of both *DMOS* and bipolar transistors. Three recurring subjects in favor of *DMOS* are *linearity, thermal stability* and *switching speed*.

Linearity

Power transistors are operated over the entire range of their output characteristic. This mode of operation is called large signal operation. *DMOS* transistors are often said to be inherently more linear than bipolar transistors because they have a roughly square law instead of exponential transfer characteristic [5,6,7]. On the other hand bipolar transistor have much higher transconductance than *DMOS* transistors. With a small resistor in the emitter, transconductance and linearity can be exchanged [8].

DMOS transistors also have a more gradual turn-on than bipolar transistors. This has been named as an important advantage in class *B* operation since a more gradual turn-on is believed to result in less crossover distortion [9,10].

Due to the different characteristics of *DMOS* and bipolar transistors they require different driving circuitry. A comparison based on transfer characteristics only is very difficult if not impossible. No decisive argument has been found in favor of either transistor type.

Thermal Stability

A more valid argument in favor of *DMOS* transistors is the excellent thermal stability. As is explained in Chapter 2, the Safe Operating Area *(SOA)* of bipolar transistors is limited by the second breakdown effect. Since the output power of an amplifier is largely determined by the ability to dissipate power in the output transistors this undeniably favors *DMOS* transistors as output transistors.

Switching Speed

Current conduction in *DMOS* transistors occurs through majority carriers only. Consequently, there is no storage of minority carriers as in bipolar transistors. This makes *DMOS* transistors very suitable for (high speed) switching applications such as the class *G* and class *D* amplifiers mentioned earlier.

In many applications the switching speed is limited by the driving circuit rather than the inherent switching speed of the power transistor. Bipolar power transistors usually have low current gain (e.g. 20) which leads to high base currents that have to be supplied by the driving circuit. DMOS transistors on the other hand have a capacitive input impedance. Although the input capacitance of DMOS power transistors can be quite large, the current that has to be supplied by the driving circuit is in general much smaller than for bipolar power transistors.

1.2.2 DMOS Output Stage

Although complementary DMOS transistors are possible [11] in most BCD technologies only n-type DMOS transistors are available. Consequently, the output stage of an audio amplifier in BCD technology has a totempole configuration as shown in Figure 1.2.1(a).

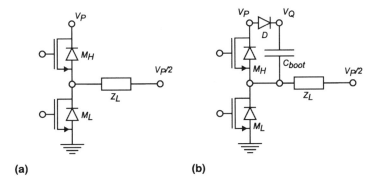

Figure 1.2.1 DMOS output stage (a) totempole structure (b) with bootstrap

The output window of this stage has a serious limitation. On the low-side, the DMOS transistor can be driven into the triode region. Consequently, the output voltage is limited by the on-resistance of the DMOS transistor which typically is in the order of 0.1Ω. For a supply voltage of 60V and a 4Ω load this results in a voltage in the order of 0.7V which is acceptable.

However, on the high-side, the output voltage is limited by the gate-source voltage of the high-side DMOS transistor and the voltage drop of the circuit that drives the gate of the high-side DMOS transistor. This voltage is at least equal to the DMOS threshold voltage which typically is in the order of 2.5V and can easily increase to 7V if the output has to source a large current. This puts an unacceptable limitation on the output voltage window.

In order to solve this problem, the gate of the high-side DMOS transistor has to be driven with a voltage V_Q that is 5V to 6V higher than the supply

Introduction

voltage V_P. Since usually only one supply voltage is available this voltage V_Q has to be generated.

One method to generate V_Q is to use a large external *bootstrap capacitor* [12] which is connected to the output of the amplifier as shown in Figure 1.2.1(b). If the output voltage is low the capacitor C_{boot} is charged through diode D. If the output voltage rises the diode D becomes reverse biased and the voltage on node V_Q is temporarily raised above the supply voltage V_P.

A second method to generate V_Q is to use a so-called *chargepump*. This is a circuit that can be integrated on-chip and requires no external components. Further, a chargepump is able to provide the required voltage V_Q continuously. In this book this latter method has been used.

1.3 Outline of this Book

In this book design aspects of monolithic audio power amplifiers in *BCD* technology are explored.

An effort has been made to acquire design knowledge that can be applied to *DMOS* power amplifiers in general. However, for the realizations described in this book, one particular *BCD* process has been used that was developed at *Philips Semiconductors*. Therefore, in the design of the actual circuits a lot of process specific knowledge has been used.

The outline of this book is as follows.

In *Chapter 2* an extensive overview is given of *DMOS* technology. The aim of this chapter is to give insight into the physical structure and operation of *DMOS* transistors in general and to describe the transistors available in the Philips *BCD* process used for the work presented in this book.

In *Chapter 3* the design of chargepump circuits is discussed. After an introduction into the principles of high voltage generation the design of a chargepump is described that is suitable for application in a *BCD* audio amplifier. The unusual ratio between the supply and chargepump output voltage and the high current that is required have led to the development of a new technique to minimize the output voltage ripple.

Chapter 4 is about the modeling of chargepump circuits. First, a general overview is given of the modeling efforts that have been published earlier. Then the development of a new improved chargepump model is described that incorporates the previous models and also includes some parasitic effects that have not been modeled before.

In *Chapter 5* the design of an audio amplifier in the Philips *BCD* process is presented. Starting from the system level, the design considerations that lead to the various amplifier circuit blocks are discussed.

Finally, in *Chapter 6* the conclusions and recommendations are given.

1.4 References

[1] Duncan, B., "Spectrally Challenged: The Top 10 Audio Power Chips", *Electronics World & Wireless World*, Vol.99, pp.804-810, Oct. 1993

[2] Moore, B.C.J. *An Introduction to the Psychology of Hearing*, 2nd Edition, Academic Press, London, 1982

[3] Benjamin, E. "Audio Power Amplifiers for Loudspeaker Loads", *Journal of the Audio Engineering Society*, Vol.42, No.9, pp.670-683, Sep. 1994.

[4] Raab. F.H., "Average Efficiency of Class-G Power Amplifiers", *IEEE Transactions on Consumer Electronics*, Vol.32, No.2, May 1986

[5] Williams, M., "Making a Linear Difference to Square Law FETs", *Electronics World & Wireless World*, Vol.100, pp.82-84, Jan. 1994

[6] Linsley Hood, J., "Expert Witness", *Electronics World & Wireless World*, Vol.101, pp.684-685, Aug. 1995

[7] Hegglun, I., "Square Law Rules in Audio Power" *Electronics World & Wireless World*, Vol.101, pp.751-756, Sep. 1995

[8] Self, D., "FETs versus BJTs; The Linear Competition", *Electronics World & Wireless World*, Vol.101, pp.387-388, May. 1995

[9] Brown, I., "Feedback and FETs in Audio Power Amplifiers", *Electronics World & Wireless World*, Vol.95, pp.123-126, Feb. 1989

[10] Brown, I., "Audio Power, FETs and Feedback", *Electronics World & Wireless World*, Vol.96, pp.343-349, Apr. 1990

[11] Botti, E., T. Mandrini, F. Stefani, "A High-Efficiency 4x20W Monolithic Audio Amplifier for Automobile Radios Using a Complementary *DMOS BCD* Technology", *ISSCC Technical Digest*, pp.386-387, 1996

[12] Brasca, G., E. Botti, "A 100V/100W Monolithic Power Audio Amplifier in Mixed Bipolar-MOS Technology", *IEEE Transactions on Consumer Electronics*, Vol.38, No.3, pp.217-222, Aug. 1992

2

DMOS Technology

The aim of this chapter is to provide some insight in various aspects of *DMOS*. After a short introduction to the role of *DMOS* in integrated circuit technology today, a qualitative overview is presented of the device physics of *DMOS*, the best known *DMOS* device structures and the technological aspects of *DMOS*. A comparison is made between *DMOS* and bipolar transistors.

2.1 Introduction

During its relatively short history, the main focus of integrated circuit technology has always been on signal processing and data storage while the field of power electronics has been dominated by the use of discrete transistors. Consequently, the process technology involved in the fabrication of power transistors has developed significantly different from that used in integrated circuits. This situation has changed since the introduction of power *MOSFET*s in the early 1970's.

2.1.1 Historical Perspective

The development of power *MOSFET*s was originally motivated by drawbacks of bipolar power transistors. Bipolar power transistors tend to have a rather low current gain which puts high demands on the driving circuitry. The switching speed of bipolar transistors is limited by the occurrence of minority carrier storage. Furthermore, the safe operating area of bipolar power transistors is limited by a destructive thermo-electrical phenomenon called *second breakdown* which will be discussed in Section 2.5.

Some of the drawbacks are eliminated if *MOSFETs* are being used. The high input resistance of *MOSFETs* makes the design of gate drive circuits easier. Since *MOSFETs* are majority carrier devices, they do not suffer from minority carrier charge storage which makes them inherently faster than bipolar transistors. The negative temperature coefficient of carrier mobility makes *MOSFETs* virtually immune to thermal problems such as thermal runaway and second breakdown. Due to these favorable characteristics, the power *MOSFETs* was first considered to be a major threat to the power bipolar transistor.

On the other hand, *MOSFETs* have some drawbacks of their own. The rather large input capacitance of *MOSFETs* turns out to be a limiting factor on the attainable speed and can contribute significantly to switching losses. The on-resistance per area is higher for *MOSFETs* while they have a lower transconductance when compared to their bipolar adversaries. A more detailed comparison between *MOS* and bipolar power transistors is presented in Section 2.5.

The process technology used in the fabrication of power *MOSFETs* is quite similar to that used in normal *MOS* and bipolar technologies. During the 1980's the high voltage integrated circuit *(HVIC)* came into the picture. In these *ICs* analog and digital functions can be integrated on the same chip as high voltage devices. This makes it possible to reduce system cost, size and weight while, in addition, the smaller number of interconnections that is needed increases reliability [1-3].

2.1.2 Introduction to DMOS

Power *MOSFET* switches have to be able to support high drain-source voltages in the off-state while in the on-state the resistance should be as low as possible in order to minimize voltage drop and thus power dissipation. Normal *MOS* transistors are not very well suited for this combination of demands as will be explained next.

The basic structure of a *NMOS* transistor is shown in Figure 2.1.1. It is composed of a *gate* electrode on a thin oxide which extends between two diffused junctions called *source* and *drain*. The source is usually connected to the bulk. The source and drain have no electrical connection unless the gate-source voltage V_{gs} exceeds the *threshold voltage* V_T which results in the creation of an inversion layer at the surface under the gate electrode.

Introduction

This inversion layer creates a conducting *channel* between source and drain. If a small drain-source voltage V_{ds} is applied then a small current will flow between source and drain. This drain current is proportional to both V_{gs} and V_{ds}. This is called the *linear region* of operation, illustrated in Figure 2.1.1(a).

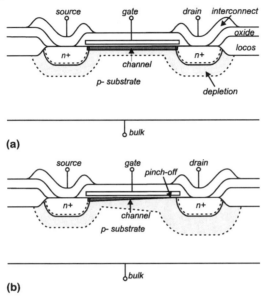

Figure 2.1.1 Basic NMOS in (a) linear region and (b) saturation region.

The drain current I_d increases when V_{ds} increases. This causes a potential gradient along the channel which leads to a decrease in the electric field towards the drain. Consequently, the inversion layer becomes thinner from source to drain and the resistance of the channel increases. When V_{ds} reaches the saturation voltage, the surface electric field near the drain is no longer strong enough to support inversion and the channel *pinch-off* occurs. If V_{ds} is increased further, most of the additional voltage drops across the depleted region between channel and drain. As a result, the depletion region increases with increasing V_{ds} and the location where the channel pinches off moves away from the drain. The resulting change in channel length is called *channel length modulation*. Since most of the additional voltage drops between channel and drain, the voltage across the channel hardly changes at all. So the drain current is now almost independent of the applied drain-source voltage. This is called the *saturation region* of operation illustrated in Figure 2.1.1(b).

In order to have a low on-resistance the channel length of a power *MOSFET* should be as small as possible. In combination with the high voltages encountered in power application this leads to rather high electric fields in the devices. This can lead to some undesirable effects. All these effects are concentrated around the drain region of the transistor.

Oxide Breakdown

If the electric field between the drain and gate exceeds a critical value then the thin oxide between them will be irreversibly damaged. Usually, the voltage across the oxide is limited to a value three or four times smaller than the critical value to prevent oxide degradation. This clearly limits the voltage handling capability of the transistor.

Punchthrough

An increase in drain-source voltage enlarges the depletion region of the drain junction. Punchthrough occurs when the depletion regions of the source and drain junction touch each other. The resulting interaction reduces the barrier for electron flow between the source and drain which results in an unwanted drain current. Punchthrough is especially significant for short channel transistors.

Avalanche Breakdown

The high electric field in the depletion region of the drain junction can lead to impact ionization. If charge carriers gain enough kinetic energy in the electric field then they create new electron-hole pairs by collision. The generated charge carriers in their turn are accelerated by the electric field and create more electron-hole pairs. This avalanche effect leads to a rapid increase in current that can damage the transistor.

Hot-carrier Injection

Charge carriers can become "hot" if their interaction with the electric field causes them to have far more kinetic energy than that corresponding to the average lattice temperature. If the kinetic energy is large enough, a hot-carrier can cross the oxide barrier and produce a gate leakage current. Trapping of hot-carriers in the oxide can lead to a permanent change in threshold voltage or transconductance of the transistor.

High voltage operation of *MOS* transistors is always achieved by including a so-called *drift region* between the drain and channel region. This drift region screens the channel from the high voltages.

Two general classes of high voltage *MOS* transistors exist. The first class is based on a redesign of the drain region of the lateral *MOSFET*s. This can be done by including a drain extension with a lower impurity dose or by using a graded impurity profile. An example of a *NMOS* transistor with an extended drain is shown in Figure 2.1.2. The lightly doped drain extension reduces the peak electric field by spreading the drain voltage over a wider depletion region. Because the doping concentration of the drain extension is usually higher than the bulk doping, the depletion region of the drain will still extend mainly into the channel resulting in substantial channel length modulation and the risk of punchthrough. This makes drain extension techniques incompatible with short channels while the improvement in voltage capability is moderate.

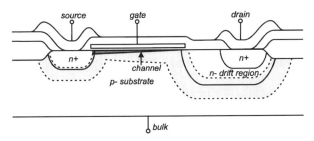

Figure 2.1.2 Extended Drain NMOS.

The second class of high voltage MOS transistors is based on DMOS technology. The 'D' in DMOS stands for *double-diffused*. The term double-diffused refers to two diffusions of complementary polarity which are subsequently implanted through the same mask opening to form the source and body of the transistor. This method is analogous to the fabrication process used to realize the emitter and base of bipolar transistors. In most DMOS processes, a self-aligning technique is used, that is, both implantations are masked on one side by a polysilicon (gate) layer.

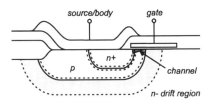

Figure 2.1.3 DMOS source-body-region.

An example of a double-diffused source-body-region for an *n*-channel DMOS transistor is shown in Figure 2.1.3. The body-region of the DMOS transistor is also sometimes referred to as *bulk region* or *backgate region* and is usually short circuited to the source diffusion.

The channel length of a DMOS transistor is determined by the difference in lateral diffusion of the source and body diffusions. If a positive gate-source voltage is applied then inversion occurs in the body-region underneath the gate similar to normal MOS transistors. The lightly doped bulk material serves as a drift region and is connected to a heavily doped drain region

DMOS Technology

which can be buried or be at the surface depending on the device structure as will be explained later.

The use of *DMOS* technology has some inherent advantages for power applications. First, the double-diffusion process enables the realization of very short channel lengths without the need for accurate lithography resulting in low channel resistance and high transconductance. The control of channel length is comparable to the control of base width in bipolar transistors. Second, the body doping is higher than the drift region doping. Consequently, if the drain-source voltage is increased then the depletion region extends mainly into the drift region. This makes the *DMOS* transistor proof against punchthrough although very short channels are used. The lightly doped drift region in *DMOS* transistors is exactly analogous to the lightly doped collector region used in bipolar transistors. [4-6]

The device physics and different device structures are discussed in more detail in Sections 2.2 and 2.3.

2.1.3 Application Areas

DMOS technology is most widely used in applications where the required breakdown voltage are relatively low (~200V) such as automotive electronics and power supplies.

In most applications, the *DMOS* transistors are used as switches. The integration of power *DMOS* transistors with low voltage devices is usually called *smart power technology*. A distinction can be made between so-called *intelligent power devices* (*IPDs*) and *power integrated circuits* (*PICs*).

An *intelligent power device* consists of one or more power *DMOS* transistors with a common drain together with other circuit elements which are used for diagnostic functions and protection of the power transistor against exceeding a current, voltage or temperature limit.

A *power integrated circuit* can contain more than one electrically isolated power *DMOS* transistors together with other low voltage circuit elements that can perform all kinds of analog and digital function [7-9].

The main difference between these smart power technologies is way the drain is contacted. In *IPDs* the drain is usually contacted at the bottom of the substrate while in *PICs* the drain contacts are located at the surface. The consequences on process technology are discussed in detail in Section 2.4.

Typical applications of *DMOS* transistors used as switches can be found in motor controllers [10,11], gas discharge lamps and battery chargers. [12,13]. Most *DMOS* transistors can also been used in analog applications such as audio [14] and video amplifiers [15], analog multiplexers [16] and line drivers [17-19].

2.2 DMOS Device Structures

With *DMOS* technology it is possible to make a number of different device structures. A distinction can be made between *lateral* and *vertical* structures. In lateral *DMOS* transistors, the heavily doped drain region is located at the surface whereas in vertical structures the drain is buried under a lightly doped epitaxial layer. A second distinction can be made in structures with a lateral channel and structures with an inclined channel.

In general, the main objective in the design of *DMOS* transistors is to minimize on-resistance for a certain breakdown voltage. Of course, on-resistance can always be reduced at the cost of more area by connecting more transistors in parallel. Consequently, the on-resistance per unit area is usually taken as a figure of merit to compare devices.

In this sections some commonly used *DMOS* structures are presented and compared. All transistors presented here are *n*-channel transistors but for *p*-channel transistors the same comparison applies.

2.2.1 Lateral DMOS

The simplest *DMOS* structure from a technological point of view is the lateral *DMOS* or *LDMOS* transistor. All terminals of an *LDMOS* transistor are contacted at the surface by the same interconnection layer. This makes the *LDMOS* a very suitable candidate for integration with other devices.

Device Structure

A cross section of an *LDMOS* transistor is shown in Figure 2.2.1.

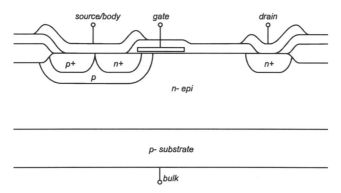

Figure 2.2.1 Cross section of a LDMOS transistor

DMOS Technology

In this transistor an epitaxial layer is used. The reversed *pn*-junction isolates the drift region from the substrate. It is also possible to realize an *LDMOS* directly into a lightly doped *n-type* substrate but this limits the application possibilities of the transistor and will not be considered here.

The breakdown voltage of an *LDMOS* transistor is in first approximation determined by the length of the drift region. This gives the designer the freedom of sizing transistors for different breakdown voltages without changing anything in the manufacturing process. In low voltage applications a self-aligned drain is sometimes used. An increase in the drift region length also results in a higher on-resistance of the transistor. In order to compensate for this a wider transistor can be used which, of course, requires more area [20-27].

Reduced Surface Fields (RESURF)

Initially, a thick epitaxial layer was chosen in order to avoid the influence of the reversed epi-substrate junction. Later it turned out that interaction between the epi-substrate depletion and the drain depletion can have a beneficial effect on the breakdown voltage due to the so-called REduced SURface Field *(RESURF)* effect [28-30].

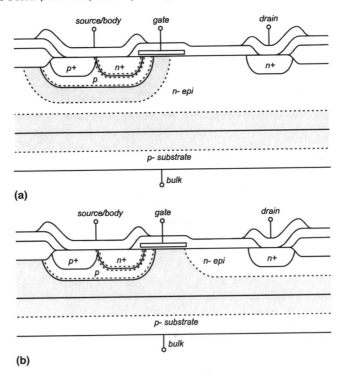

Figure 2.2.2 LDMOS depletion (a) without RESURF and (b) with RESURF

DMOS Device Structures

The interaction causes a horizontal stretch of the drift region depletion. The applied drain voltage is then distributed over a longer lateral distance which reduces the peak electric field. This effect is illustrated in Figure 2.2.2. This makes it possible to use thinner epitaxial layers while still achieving high breakdown voltages.

Current Flow

In an *LDMOS* transistor, the current always flows close to the surface. It passes through three regions from source to drain as indicated in Figure 2.2.3.

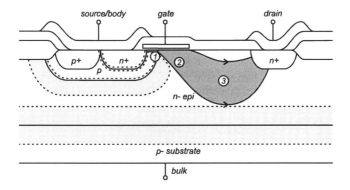

Figure 2.2.3 Current flow in a LDMOS transistor

In region *1*, the channel is formed by a thin inversion layer near the surface of the transistor. In region *2*, directly after the channel, the gate overlaps the drift region. When the transistor is in the on-state, this region directly under the gate is heavily accumulated. Most of the current flows through this accumulation region while the remaining current spreads into the drift region below. In region *3* the current spreads further into the bulk until it reaches the drain region [22,31].

On-resistance

All three regions described contribute to the total on-resistance of the transistor. If the transistor is turned on strongly then the channel and accumulation regions have a negligible contribution to the on-resistance and the on-resistance is almost entirely determined by the drift region. In *RESURF* devices, the effective thickness of the drift region is modulated by the depletion region of the epi-substrate junction which acts as a *JFET* [21,32-34].

Breakdown

Breakdown of an *LDMOS* transistor can occur at several places. Punchthrough can occur between source and drain through the channel or, if the transistor is used in a high-side configuration between source and substrate through the epi-layer. Avalanche breakdown can occur at the reversed body-epi junction or, if present, the epi-substrate junction. The breakdown voltage of the epi-substrate junction is normally the highest and defines the upper limit of the transistors breakdown voltage.

The body-epi junction is the most likely spot for breakdown because of the curvature of the depletion which leads to concentration of electric field lines. Several techniques can be used to reduce this curvature. An extension of the source metallization acting as a field plate helps to deplete the surface and to reduce the curvature. The potential difference between the drift region and the field plate effectively pushes the current down into the substrate. If the transistor is only used in low-side application a buried *p*-layer under the body-region can be used to increase the depletion from the substrate side. For a transistor in high-side application this may lead to punchthrough. In that case depletion can be enhanced from the top side by using a floating *p-type field ring*. These methods are illustrated in Figure 2.2.4.

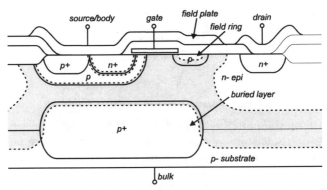

Figure 2.2.4 Enhancement of the breakdown voltage

Drain-source breakdown voltages as high as *1200V* have been reported using *LDMOS* transistors [26,32,33,35-37].

Equivalent Circuit

An equivalent circuit of a *LDMOS* transistor is shown in Figure 2.2.5. The channel region is represented by an enhancement *MOS* transistor. The accumulation region is modeled by a depletion *MOS* transistor in parallel with a resistor. The remainder of the drift region is modeled with a second resistor. In *RESURF* transistors, both resistors are modulated by the epi-substrate junction. The diode represents the body-epi junction [20].

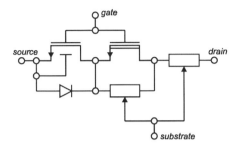

Figure 2.2.5 Equivalent circuit of a LDMOS transistor

Layout

The layout of a *LDMOS* transistor usually has a finger structure similar to that used in layout of normal *MOS* transistors. However, if the *LDMOS* transistor is realized without an epitaxial layer, care has to be taken that the drain is entirely surrounded by the source-body-region in order to isolate it from the periphery [21].

2.2.2 Vertical DMOS

The most widely used *DMOS* structure is the vertical *DMOS* or *VDMOS*. The drain of a *VDMOS* is a buried diffusion underneath a lightly doped epitaxial layer which acts as a drift region. In discrete *VDMOS* transistors the drain is formed by the substrate and is usually contacted at the bottom side of the transistor. If integration with other components is required then the buried drain has to be contacted from the surface with deep sinker diffusions.

Device Structure

The source and body-region of a *VDMOS* transistor is usually constructed as an array of many small symmetrical cells connected in parallel and separated by the gate electrode that surrounds each cell. The gate width is determined by the perimeter of one cell times the number of cells.

A cross section of a typical *VDMOS* transistor is shown in Figure 2.2.6. The diffusion profile of a *VDMOS* transistor is similar to that of a vertical bipolar transistor. Actually, the same diffusions can be used to produce a pretty good bipolar transistor as well. The attainable breakdown voltage is limited by the thickness and doping concentration of the epitaxial layer and can not be changed by the designer. The designer still has the possibility to change the spacing between neighboring cells which also has a strong influence on both breakdown voltage and on-resistance.

DMOS Technology

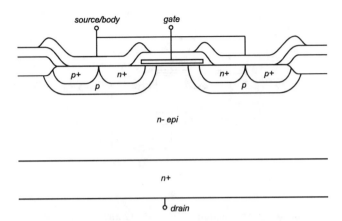

Figure 2.2.6 Cross section of a VDMOS transistor

Between adjacent cells the drift region is covered completely by the gate electrode which results in a accumulation layer whenever a positive gate bias is applied [20,25,27].

Current Flow

Contrary to LDMOS transistors, the current in VDMOS transistors flows near the surface as well as in the bulk. Five regions of current flow can be distinguished as indicated in Figure 2.2.7.

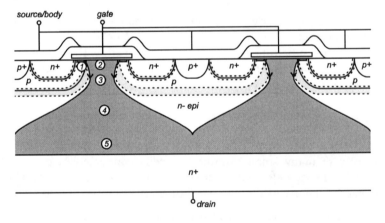

Figure 2.2.7 Current flow in a VDMOS transistor

In region 1, similar to LDMOS the channel is formed by a thin inversion layer near the surface. In region 2, the accumulation layer is entered which is also near the surface. Along the accumulation layer the current gradually changes direction and flows down towards the buried drain. in region 3, the current passes the region between adjacent cells which is pinched by the depletion regions of the body-epi junctions. This region is often called *JFET* or *neck region*. In region *4*, after leaving the neck region the current spreads in the drift region in a trapezoidal manner. Finally, in region *5*, the currents from adjacent regions meet and current flows homogenically further down to the buried drain region [20,38].

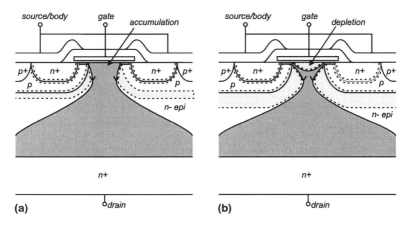

Figure 2.2.8 Current flow at (a) moderate and (b) high drain voltages.

Quasi-saturation

The drain voltage is of particular influence on the current profile in the neck region. For this reason the current flow in the accumulation and neck region are treated in more detail here.

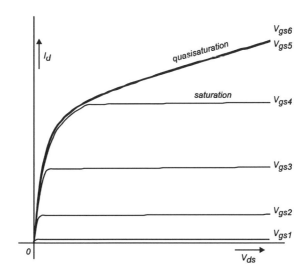

Figure 2.2.9 Effect of quasi-saturation on drain current

If moderate drain voltages are applied then the surface under the gate electrode is completely accumulated and the current spreads homogeneously downward as shown in Figure 2.2.8(a). At higher drain voltages, a depletion region is formed underneath the gate electrode while at the same time the depletion region of the body-epi junction extends. The current is now concentrated in a narrow path between these depletion regions as shown in Figure 2.2.8(b). Due to the high electric field the electron drift velocity saturates which, in turn causes the current to saturate. This phenomenon is called *quasi-saturation* [39-44].

The quasi-saturation effect in *VDMOS* transistors causes a low output resistance and insensitivity to increase in the gate voltage at high current levels as shown in Figure 2.2.9.

On-resistance

Similar to *LDMOS* transistors, all regions of current flow contribute to the total on-resistance. If the transistor is turned on strongly, the influence of the channel and the accumulation region can be neglected and the neck and drift regions dominate the on-resistance. The contribution of the neck region and the trapezoidal part of the drift region are strongly influenced by the size of and the spacing between the source-body cells. First, as can be seen in Figure 2.2.7, the region directly under the source-body cells does not conduct any current. The volume of this region should be minimized in order to decrease on-resistance per unit area. Consequently, the source-body cells should be made as small as possible. Second, the width of the neck region can be increased by increasing the spacing between cells. A lower limit of the on-resistance is defined by the thickness and resistivity of the epi-layer [45-50].

Breakdown

Breakdown in a *VDMOS* transistor can occur due to punchthrough between source and drain through the channel or due to avalanche breakdown of the body-epi junction or, if present, of the drain-substrate junction. Similar to *LDMOS* the body-epi junction is most susceptible to breakdown due to the curvature of the depletion.

In the region between neighboring source-body cells the curvature can be reduced significantly by reducing the cell spacing. This causes the depletion regions of the body-epi junction to interact and to smoothen the depletion surface as illustrated in Figure 2.2.10. This effect is known as the *proximity effect*. The field plate action of the gate electrode also has a beneficial effect on the curvature. At the boundaries of a *VDMOS* transistor other techniques have to be used to reduce the junction curvature. Some well known techniques are summarized in Section 2.4.

DMOS Device Structures

Apparently, reduction of on-resistance and increase of breakdown voltage lead to conflicting demands on cell spacing and a trade-off has to be made.

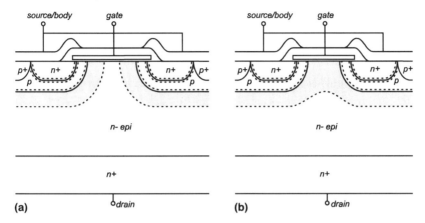

Figure 2.2.10 Influence of cell spacing on depletion

An upper limit for the breakdown voltage is determined by the thickness and doping of the epitaxial layer. The maximum epitaxial thickness that is economically feasible in integrated circuits is about *25μm*. This limits the drain-source breakdown voltage to approximately *250V* for vertical transistors [7,51-54].

Equivalent Circuit

An equivalent circuit of a *VDMOS* transistor is shown in Figure 2.2.11.

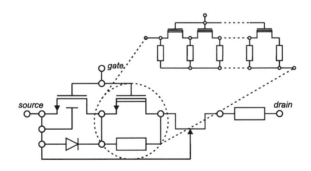

Figure 2.2.11 Equivalent circuit of a VDMOS transistor

Similar to *LDMOS* transistors, the channel region is represented by an enhancement *MOS* transistor. The accumulation region is modeled by a parallel connection of a depletion *MOS* transistor and a resistor. These components are best modeled as distributed transistors to account for the

DMOS Technology

gradual change in direction of the current along the accumulation layer. The neck region is represented by a *JFET* and the rest of the drift region by a simple resistor. Contrary to *LDMOS* transistors, this resistance is not influenced by the substrate due to the buried drain. The diode represents the body-epi junction [20].

Layout

As mentioned earlier, the source-body-region is normally constructed as many symmetrical cells that are separated by the surrounding gate electrode. Several polygon shapes, for example hexagonal or square, are being used but no particular geometry has a major advantage with regard to optimizing on-resistance and breakdown voltage.

If a *VDMOS* transistor needs a drain contact at the surface then the drain has to be brought up to the surface periodically with deep sinker diffusions. The buried drain sheet resistance and sinker resistance add to the total on-resistance. Care has to be taken to determine the optimal number of cells between sinker diffusions in order to minimize the on-resistance per unit area [55].

2.2.3 VMOS and UMOS

A rather exotic *DMOS* structure is the *V-groove MOS* or *VMOS* transistor. In contrast to *LDMOS* and *VDMOS* transistors presented previously, the channel of a *VMOS* transistor is at the surface but located along the sides of an anisotropically etched groove [56].

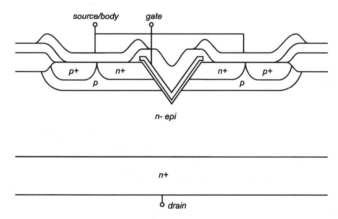

Figure 2.2.12 Cross section of a VMOS transistor

While *LDMOS* and *VDMOS* transistors can be realized on any crystalline orientation, *VMOS* transistors are constrained to have the channel along an etched <111> plane. *VMOS* transistors are commonly realized as vertical transistors although a lateral structure is also possible. Due to the specific processing steps necessary to realize the *V-groove*, *VMOS* transistors are hardly ever combined with other devices.

Device Structure

Apart from the *V-groove*, the structure of a *VMOS* transistor is very similar to a *VDMOS* transistor. A cross section of a *VMOS* transistor is shown in Figure 2.2.12. Since the channel is not parallel to the surface but is inclined, the channel length is determined by the difference in vertical rather than lateral diffusion depth. The angle of the channel with the surface is *54.7°*. For the same diffusion schedule, the channel of a *VMOS* is about *1.5* times longer than that of an *LDMOS* or *VDMOS* transistor [20,57].

Current Flow

The current flow in a *VMOS* transistor has similarities to that of both *LDMOS* and *VDMOS* transistors. Four regions can be distinguished as indicated in Figure 2.2.13.

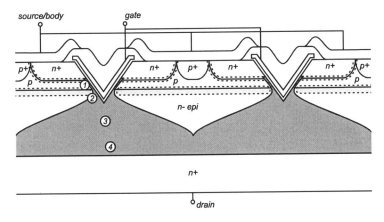

Figure 2.2.13 Current flow in a VMOS transistor

In region *1*, the channel is formed by a thin inversion layer near the *V-groove* side. In region *2*, after the channel region, an accumulation layer is formed where the current starts spreading into the drift region towards the drain. These two regions resemble the first two regions in an *LDMOS* transistor. In region *3*, at the bottom of the *V-groove*, the accumulation region ends and the current spreads further in a trapeziodal manner. In region *4*, the current from adjacent *V-grooves* meet and current flows homogenically down to the drain. These last two regions are identical to the last two regions in a *VDMOS* transistor [20].

On-resistance

The resistance of the channel and accumulation regions in VMOS transistor is higher than in LDMOS and VDMOS transistors. This has to do with the crystalline orientation of the silicon surface under the gate electrode. In LDMOS and VDMOS transistors the channel is along the <100> plane while the VMOS channel is etched along the <111> plane which results in a lower electron mobility in both the inversion and accumulation layers.

A typical difference between VMOS and VDMOS transistors is the absence of a neck region in VMOS. Therefore the on-resistance of VMOS can be lower than for VDMOS when the same epitaxial thickness and doping are being used. The on-resistance can be reduced by making a deeper V-groove in order to enlarge the accumulation region. However, this increases the area occupied by the groove and reduces the breakdown voltage [20,45].

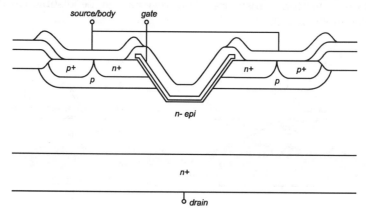

Figure 2.2.14 Cross section of a UMOS transistor

Breakdown

Breakdown in a VMOS transistor can, similar to LDMOS and VDMOS transistors, be caused by punchthrough between source and drain through the channel or by avalanche breakdown of the body-epi junction. The sharp apex of the V-groove results in very high electric fields which lead to much lower breakdown voltages than VDMOS transistors with comparable dimensions. It turns out that for the same breakdown voltage, VMOS has a higher on-resistance than VDMOS. An improvement of the electric field distribution can be realized by truncating the V-groove. This leads to a U-groove MOS or UMOS as shown in Figure 2.2.14. Besides reducing the electric field, the accumulation layer at the bottom of the U-groove enhances distribution of current in the drift region and thus reduces on-resistance [58].

A drawback of the UMOS structure is that much better process control is required compared to the self-stopping V-groove process.

Equivalent Circuit

The equivalent circuit of a VMOS transistor is shown in Figure 2.2.15. It is essentially identical to that of the VDMOS transistor without the JFET.

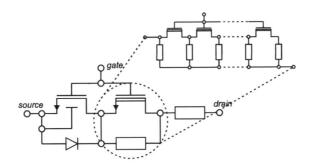

Figure 2.2.15 Equivalent circuit of a VMOS transistor

Layout

The layout of VMOS and UMOS transistors usually has a finger structure similar to LDMOS transistors. It is also possible to use an array of cells similar to VDMOS transistors. However, this is not very attractive since it results in a lower channel width per unit area.

2.2.4 Comparison of DMOS Structures

In case integration with other components is required, the LDMOS and VDMOS structures are more suitable than VMOS since they can be realized using standard manufacturing steps.

The LDMOS structure has the additional advantage that it has all electrodes at the surface without the need for area consuming sinker diffusions. Further, LDMOS is capable of very high breakdown voltage without the need for thick epitaxial layers.

Up to a certain breakdown voltage VDMOS transistors have lower on-resistance per unit area than LDMOStransistors due to the more efficient use of drift region volume. In many smart power applications it is no problem to contact the drain at the bottom side. In that case VDMOS is favorable [20,32].

2.3 DMOS Device Properties

Although the operation of DMOS transistors is basically similar to that of standard MOS transistors, its behavior is only crudely represented by standard MOS models. The main differences are caused by the very short and nonuniformly doped channel region and the lightly doped drift region in series with the channel. In this section some of the properties of DMOS transistors relevant to circuit design are considered.

2.3.1 Transfer Characteristics

As mentioned earlier, the channel region of a DMOS is realized using a double-diffusion process. The resulting doping profile along the channel shown in Figure 2.3.1 is similar to that in the base of a bipolar transistor.

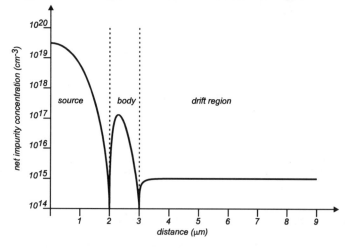

Figure 2.3.1 Typical doping profile along a DMOS channel

The doping concentration is highest at the source side and decreases towards the drain [59,60].

Threshold Voltage

The threshold voltage of normal MOS transistors is related to the doping concentration in the channel. Since the channel doping in DMOS decreases towards the drain, the threshold voltage also decreases towards the drain.

If the gate-source voltage is increased at low drain voltages then the inversion of the channel region starts from the drain side and then gradually grows towards the source side. This gradual inversion mechanism results in a rather long transition from subthreshold to strong inversion. Current flow between source and drain is possible as soon as the inversion layer reaches the source diffusion. Therefore, the effective threshold voltage in *DMOS* is determined by the highest doping level in the channel. Due to the high channel doping, the value of the threshold voltage in *DMOS* is usually higher than for normal *MOS* transistors.

Since the body and source diffusions are always short circuited, the threshold voltage in *DMOS* transistors is not influenced by the body effect.

The diffusion dopants are commonly implanted at a 7° angle to the normal of the surface. This can result in differences in peak channel doping depending on the orientation of the channel resulting in significant threshold mismatch [18,22,45,61-63].

Transconductance

In contrast to normal *MOS* transistors which have a more or less square law transfer characteristic, the transfer characteristic of *DMOS* transistors is more linear especially at higher drain bias. This linear behavior is caused by three effects.

First, due to the short channel of *DMOS* the electric field in the channel becomes very high. The high electric field causes the electrons in the channel to reach their scattering limited velocity. This effect also occurs in normal *MOS* transistors with short channels.

Second, one of the effects caused by the graded doping profile of the channel is that the gain factor, which is assumed to be constant in normal *MOS* transistors, depends on the applied gate bias. This is a result of the graded doping profile of the channel of *DMOS*. The enhancement type *MOS* transistor that is often used to model the channel region of *DMOS* transistors can be considered as a series configuration of a large number of *MOS* transistors with different threshold voltages and gain factors. The threshold voltage decreases from source to drain while the gain factor increases in this direction due to the higher electron mobility in regions with lower doping. The resulting effective gain factor decreases with increasing gate bias.

Third, the drift region located between the drain and channel causes the voltage across the channel to decrease with increasing drain current. This has a linearizing effect on the transfer characteristic.

The less than square law transfer characteristic of *DMOS* transistors has been mentioned as an advantage compared to the exponential transfer characteristic of bipolar transistors because it results in more linear voltage to current conversion [33,60,64-67].

Output Resistance

Due to the lightly doped drift region and heavily doped body-region, the depletion of the reverse biased body-drift junction extends mainly into the

DMOS Technology

drift region. Consequently, channel length modulation in *DMOS* transistors is small even though very short channels are used. This results in a very high output resistance [59].

2.3.2 Parasitic Elements

Inherent to the structure of *DMOS* transistors are a number of parasitic elements that should be taken into account in circuit design.

Source Drain Diode

In *DMOS* transistors, the source and body-region are short circuited by the source metallization. As a result the *pn*-junction formed by the body-region and the drift region, which is reverse biased in normal operation, introduces a parasitic diode between the source and the drain. This diode is included into the circuit symbol for *DMOS* as shown in Figure 2.3.2.

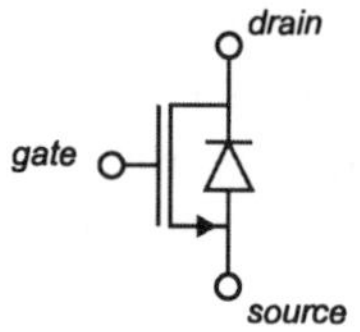

Figure 2.3.2 DMOS Circuit Symbol

The result of this diode is that a *DMOS* transistor can only block voltages in one direction. If the drain voltage drops below the source voltage the diode starts to conduct and gate control is lost.

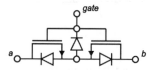

Figure 2.3.3 Bidirectional DMOS switch

The parasitic diode can be exploited in rectifier circuits. The voltage drop of the forward biased diode can be reduced by turning on the *DMOS* channel which then acts as a shunt resistance for the diode. A bidirectional switch can be realized by connecting two *DMOS* transistors back to back as shown in Figure 2.3.3 [68-70].

Input Capacitance

The input capacitance of DMOS transistors plays an important role in the design of driver circuits. In high frequency switching applications, the gate charge and discharge currents can give a dominant contribution to the power dissipation. The input capacitance consists of the gate-source and gate-drain capacitance as is shown in Figure 2.3.4.

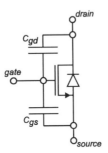

Figure 2.3.4 Input capacitance

The gate-source capacitance has two parts. First, all DMOS transistors inherently have some overlap of the gate electrode over the heavily doped source diffusion. Especially in VMOS transistors this overlap is relatively large. The resulting capacitance is independent of the applied gate-source voltage. Second, the gate-channel capacitance present in all MOS transistors contributes to the gate-source capacitance.

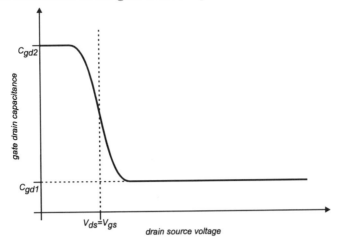

Figure 2.3.5 Gate-drain capacitance

DMOS Technology

This capacitance depends on the applied gate-source voltage but can be considered constant if the gate-source voltage exceeds the threshold voltage.

A significant contribution to the input capacitance can come from the gate-drain capacitance. This capacitance is much larger than in normal MOS transistors due to the relatively large overlap of the gate electrode on the drift region and depends on the gate-drain voltage. This dependence is relatively strong in VDMOS transistors. As was shown earlier in Figure 2.2.8 the accumulation near the surface becomes depleted when high drain voltages are applied. This results in a change in gate-drain capacitance that can reach an order of magnitude. The variation of gate-drain capacitance with drain-source voltage is shown in Figure 2.3.5.

Some efforts have been reported to reduce the drain gate capacitance by selectively removing the gate or using a thicker gate oxide at the drift region overlap [50,60,65,71-74].

Parasitic Bipolar Transistor

The diffusion profile of DMOS transistors is very similar to that of bipolar transistors. A direct result of this similarity is a parasitic bipolar transistor between source and drain which is inherent to the DMOS structure. The body-region is the base of this transistor.

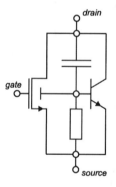

Figure 2.3.6 Turn-on of the parasitic bipolar transistor

At the surface, the base/emitter (source/body) is shorted and forms the source-drain diode. However, it is possible that the bipolar transistor is turned on during high speed turn-off transients of the DMOS transistor. This is due to the capacitance between bulk and drain and the finite resistance of the body-region under the source. An equivalent circuit is shown in Figure 2.3.6. If the drain voltage rises fast enough, the voltage drop over the buried body resistor can become high enough to forward bias the base-emitter junction and to turn on the bipolar transistor which may lead to the destruction of the transistor. The impact of the parasitic bipolar transistor on the breakdown behavior and reliability of *DMOS* transistors is discussed in more detail in Section 2.5. [45,50]

2.4 DMOS Process Technology

High voltage integrated circuits emerged in the 1980's. In standard *MOS* technology the major interest in technology improvement has always been reduction of the minimum feature size in order to reduce the price per transistor. In case high voltage devices are included in integrated circuits, other aspects gain more importance such as the isolation between devices and the termination at the boundaries of the high voltage devices required to increase breakdown voltages. In this section these technological aspects of *DMOS* processes are discussed.

2.4.1 Isolation Techniques

Isolation between high voltage devices and reduction of cross talk to sensitive low power parts are more dominant issues in high voltage integrated circuits than in standard integrated circuits. This is due to the much higher signal levels encountered in high voltage applications. High voltage technologies can be classified into three categories according to the isolation technique that is used.

Self Isolation

The simplest isolation technique is self isolation. In self isolating technologies the reverse biased junctions of source and drain diffusions provide the isolation. Self isolation is a commonly used technique in *CMOS* technologies. Two examples of self isolated high voltage technology are shown in Figure 2.4.1.

In Figure 2.4.1(a) a *p-type* substrate is used as a body for the *LDMOS* and *NMOS* transistors while for the *PMOS* an *n*-well is used. The lightly doped drift region of the *LDMOS* requires an additional *n-diffusion*. In this type of technology, an annular layout has to be used for the *LDMOS* in which the drain is completely surrounded by the source which is always grounded. Furthermore, this technology is limited to lateral transistors.

A self isolating technology with a vertical *DMOS* structure is shown in Figure 2.4.1(b). Here the *PMOS* can be realized directly into the *n-type* drift region/substrate while for the *NMOS* a *p*-well is used. The doping of the substrate is high in order to get a low drain resistance. A drawback of this technology is that *DMOS* transistors always have a common drain.

DMOS Technology

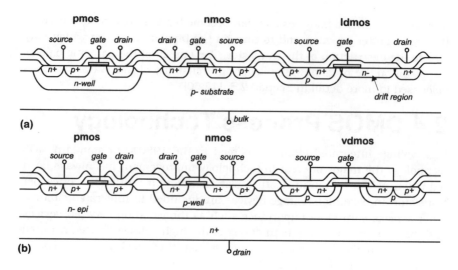

Figure 2.4.1 Self isolation

Simply stated, the technology in Figure 2.4.1(a) is the result of adding a high voltage transistor to a standard *CMOS* process whereas the technology in Figure 2.4.1(b) is the result of adding *CMOS* to a standard discrete power transistor process. Apparently, self isolation precludes complete isolation between high voltage devices [2,9,21].

Junction Isolation

In junction isolated processes the reverse biased junction between an epitaxial layer and the substrate is used to provide isolation. Actual isolation between devices is realized with deep sinker diffusions. An example of a junction isolated technology is shown in Figure 2.4.2.

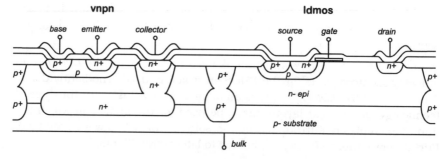

Figure 2.4.2 Junction isolation

An advantage of junction isolation is that multiple high voltage transistors can be integrated in all possible configurations. Junction isolation is common practice in most bipolar and *BiCMOS* technologies. In fact, many

DMOS technologies are based on *BiCMOS* technologies. A disadvantage of junction isolation is the rather large area that is usually required by the lateral isolation [2].

Dielectric Isolation

A disadvantage of both self isolation and junction isolationis the sensitivity to latch-up effects due to the omnipresent parasitic thyristor structures. A very effective but also expensive isolation technique is dielectric isolation. In dielectric isolated processes, devices are isolated from one another with layers of insulating material, usually silicon oxide. Two examples of a dielectric isolated technology are shown in Figure 2.4.3.

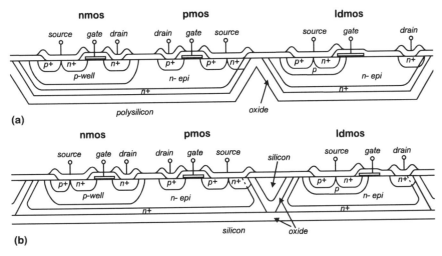

Figure 2.4.3 Dielectric isolation (a) polysilicon deposition (b) direct bonding

In the technology shown in Figure 2.4.3(a) the isolated tubs are realized by a deep *V-groove* etch followed by thermal oxidation. Then polysilicon is deposited for mechanical support after which the surface is ground and polished to the required thickness. In the technology shown in Figure 2.4.3(b) direct bonding is used to connect two oxidized wafers followed by grinding and polishing of surface to the required thickness. Then a deep *V-groove* etch followed by thermal oxidation is used to realize lateral isolation.

The area needed for dielectric isolation is much smaller than for junction isolation resulting in high packing densities.

A possible disadvantage of dielectric isolation is a high thermal resistance between the power transistors and substrate due to the low thermal conductivity of the silicon oxide which can lead to excessive self-heating. The thickness of the oxide layer between the transistor and the substrate has a significant influence [2,16,75-77].

2.4.2 Device Termination Techniques

In general, breakdown in the lateral direction of a pn-junction occurs well below the ideal vertical breakdown limit due to increased electric fields at the edges of the junction. The increased electric fields are a result of the junction curvature. In order to approach the ideal breakdown limit, special junction termination techniques have to be used that spread out the depletion region and thus reduce the peak electric field at the surface [45,58,70,78-80].

Floating Field Ring

A floating field ring is an electrically floating junction that is placed near the edge of the planar junction. The depletion region of the floating field ring enhances the depletion of the planar junction as shown in Figure 2.4.4(a). If the spacing between the floating field ring and the planar junction is optimized, breakdown occurs at both junctions simultaneously. The floating field rings can be realized with the same diffusion as the pn-junction. Breakdown voltages up to 80% of the ideal vertical limit can be achieved with this technique [27,78].

Field Plate

A field plate is an electrode on a dielectric layer that overlaps the planar junction edge. The surface potential can be influenced by applying the appropriate potential to the field plate as shown in Figure 2.4.4(b). An ideal field plate structure has a thin dielectric layer near the junction edge that thickens away from the junction. Field plates are often used in combination with floating field rings. The use of field plates results in about 60% of ideal breakdown voltage [27,78,80].

Junction Termination Extension

One or more regions of decreasing doping concentration can be attached to the planar junction edge resulting in a junction termination extension. The effect of this technique resembles a graded junction and results in an extended depletion region as shown in Figure 2.4.4(c). Up to 95% of the ideal breakdown voltage can be achieved with this technique [58,78,81].

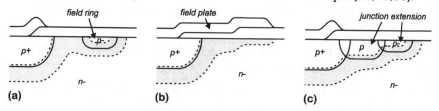

Figure 2.4.4 Device termination techniques (a) floating field ring (b) field plate (c) junction termination extension

2.4.3 BCD Technology

High voltage integrated circuit technologies that combine Bipolar, *CMOS* and *DMOS* transistors are called *BCD* technologies. Many *BCD* technologies are based on Bipolar or *BiCMOS* technologies to which a *DMOS* transistor is added. Optimization of the *DMOS* transistors generally results in a degeneration of the performance of the other components. In most *BCD* technologies, only *n*-channel *DMOS* transistors are available. Integration of complementary *DMOS* transistors is more expensive while the *p*-channel transistors require extra mask steps and about three times more area for the same on-resistance compared to the *n*-channel transistors. A typical single-metal single-poly junction isolated *BCD* process involves between *10* and *12* masking steps. A short description of the process flow and the devices of the *BCD* process used for the designs in this book is presented next [25,27,60,75,82-86].

Process Flow

A cross section of the most important devices at some key steps in manufacturing is shown in Figure 2.4.5. Starting material is a lightly doped *p-type* substrate.

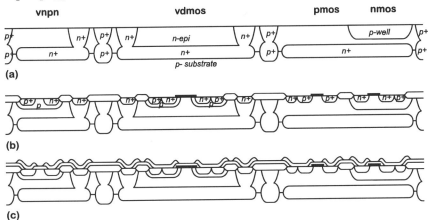

Figure 2.4.5 Key steps in BCD process flow

I. Implantation of a Buried *N+* (*BN*) layer to define the buried drain for *VDMOS* and buried collector for vertical *NPN*. Often, *BN* is also placed underneath shallow *p-type* diffusions to prevent punchthrough to the substrate.

II. Implantation of a Buried *P+* (*BP*) layer for isolation.

III. Epitaxial growth of about *8µm* lightly doped *n*-layer and diffusion of the *BN* and *BP* implantations.

IV. Implantation of Deep N+ (DN) sinker diffusions to contact buried drain and collector regions.

V. Implantation of Deep P+ (DP) sinker diffusions to close junction isolation.

VI. Implantation of P Well (PW) for low voltage NMOS. The cross section shown in Figure 2.4.5(a) corresponds to the steps up to this moment.

VII. Definition of the oxide (OD) and growth of LOCOS[1].

VIII. Growth of 70nm gate oxide and PolySilicon deposition (PS) and pattern definition for the gate electrodes of DMOS and CMOS transistors.

IX. Implantation of Shallow P (SP) to form the body-region of DMOS and base region of vertical NPN.

X. Implantation of heavily doped Shallow P (SP2) for source and drain regions of low voltage PMOS and ohmic contact to Shallow P and P-Well diffusions.

XI. Implantation of heavily doped Shallow N (SN) for source and drain regions of low voltage NMOS and DMOS and emitter region of vertical NPN. The cross section shown in Figure 2.4.5(b) corresponds to the steps up to this moment.

XII. Definition of contact openings (CO) and deposition of TEOS[2] oxide.

XIII. Interconnect metal deposition (IN). The cross section shown in Figure 2.4.5(c) corresponds to this moment.

XIV. If required, a second interconnect can be added which results in two additional masking steps (CO2 and IN2).

XV. Deposition of passivation layer and definition of the bonding openings

VDMOS

A cross section of a VDMOS transistor is shown in Figure 2.4.6. As explained earlier, VDMOS transistors are comprised of an array of cells connected in parallel. The corner cells are inactive in order to prevent premature breakdown. Consequently, the smallest transistor has 3x3 cells of which 5 are active and has a gate width of about 250µm and a cell pitch of 22µm. Large power transistors are made of fingers of 7x30 cells enclosed in comb-shaped drain sinkers.

---

[1] localized oxidation of silicon

[2] tetraethylorthosilicate: $Si(C_2H_5O)_4 \rightarrow SiO_2$ + byproducts

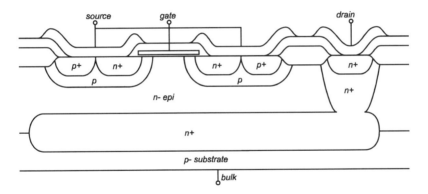

Figure 2.4.6 VDMOS transistor

In analog designs special attention should be paid to the large variations in input capacitance depending on the drain bias as was discussed in Section 2.3.

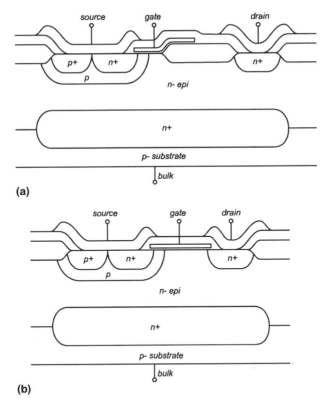

Figure 2.4.7 LDMOS transistor for (a) high voltage and (b) low voltage

DMOS Technology

LDMOS

For *LDMOS* transistors a high voltage and a low voltage configuration can be used. In the high voltage configuration, the oxide underneath the gate changes from thin to thick oxide as shown in Figure 2.4.7(a). This is done to reduce field crowding at the gate edge and to protect the thin oxide against high drain voltages.

In the low voltage configuration the drain is self-aligning as shown in Figure 2.4.7(b). This results in a much lower on-resistance because the entire drift region can be accumulated. The *BN* layer is meant to prevent premature punchthrough between the body-region and substrate. The breakdown voltage of the high voltage *DMOS* transistors is about *70V*.

EPMOS

The extended drain *PMOS* transistor is the only transistor available with the same voltage capability as the *DMOS* transistors. A cross section is shown in Figure 2.4.8.

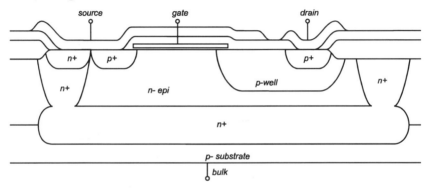

Figure 2.4.8 Extended drain PMOS transistor

The entire transistor is put inside a *BN/DN* tub which is connected to the source. The drain extension is made with *PW*. Although *PW* has the lowest doping of the available *p-type* diffusions it is still to high for sufficient *JFET* action. Therefore, the drift region is implemented as a pattern of small stripes to reduce the average doping concentration. A drawback of this somewhat provisional structure is that the desired *JFET* action begins at relatively high drain-source voltages which results in a sudden increase in output resistance.

The channel length must be chosen rather long due to the overlap of the gate electrode on the drift region and considerable channel length modulation. As a result, *EPMOS* transistors are usually quite large compared to other transistors.

CMOS

A cross section of the CMOS transistors is shown in Figure 2.4.9. In contrast to the other transistors which are isolated individually, CMOS transistors can be combined in one isolated region thanks to the self-isolated nature of the CMOS structure.

If CMOS is used at voltages higher than 5V then it is necessary to surround the transistor with a so-called channelstopper to prevent parasitic channels between transistors. This can be done with a ring of SN for the PMOS and SP2 for the NMOS transistors.

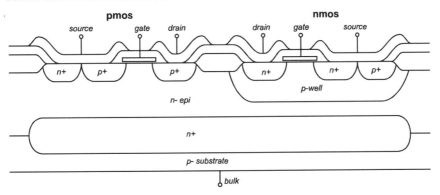

Figure 2.4.9 CMOS transistors

VNPN

A cross section of a vertical NPN transistor is shown in Figure 2.4.10. The diffusion profile is identical to that of the VDMOS. However, the open-base breakdown voltage is about 18V which is considerably lower than the DMOS breakdown voltage.

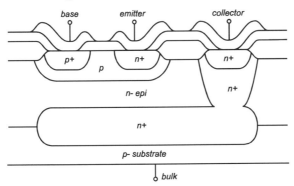

Figure 2.4.10 Vertical NPN transistor

DMOS Technology

The *VNPN* transistors have a current gain of about *60* which is almost constant over a wide range of the collector current density.

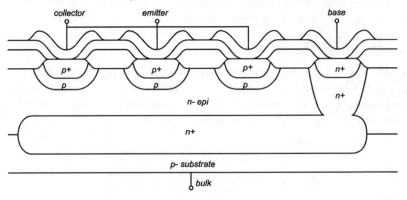

Figure 2.4.11 Lateral PNP transistor

LPNP

A cross section of a lateral *PNP* transistor is shown in Figure 2.4.11. The collector diffusion is an octagonal ring around the emitter region. The transistor action takes place at or near to the surface. Together with the unfavorable diffusion profile the *PNP* performance is inferior to that of the *VNPN*. Due to the structure of the *LPNP* the base-substrate capacitance is considerable. The open-base breakdown voltage of the *LPNP* transistors is also limited to about *18V*. The current gain ranges from *10* to *100* depending on the collector current density.

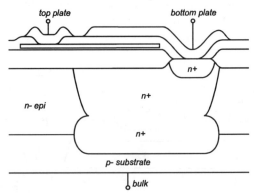

Figure 2.4.12 High voltage sandwich capacitor

DMOS Process Technology

Capacitors

Several capacitor structures are possible. Low voltage capacitors can be made with *PS* as one plate, epi, *PW* or *DN* as the other plate and gate oxide as a dielectric. For high voltages the gate oxide cannot be used and the *LOCOS* and *TEOS* oxides have to serve as a dielectric. A cross section of the resulting sandwich structure is shown in Figure 2.4.12.

Diodes

Some cross sections of possible diodes are shown in Figure 2.4.13. The *SP-epi* diode shown in Figure 2.4.13 (a) is similar to the parasitic source-drain diode of the *DMOS* transistors.

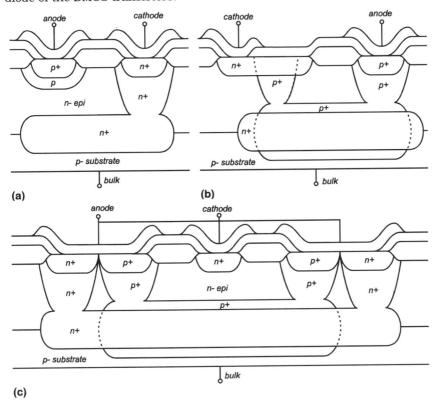

Figure 2.4.13 Diode structures: (a) SP-epi (b) Zener (c) Low substrate leakage

The diode in Figure 2.4.13(b) uses a combined BN/BP layer to shield the diode from the substrate. Here, the diode is formed by the epi-BP/DP junction. The diode in Figure 2.4.13(c) uses the heavily doped SN and DP diffusions to obtain a Zener diode.

Resistors

Both diffusion and poly resistors can be used. Poly resistors are more linear than diffusion resistors but practical values are limited to about 10kΩ. For higher values a *SP* or pinched *SP* resistor can be used. Particularly the pinched *SP* resistor shown in Figure 2.4.14 has a strong dependence on the applied voltage due to the *JFET* like operation.

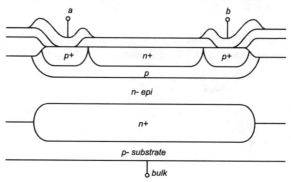

Figure 2.4.14 Pinched SP resistor

2.5 DMOS Versus Bipolar

Probably the most decisive reason to use *DMOS* instead of bipolar transistors in power applications is reliability. A familiar problem in bipolar transistors is the occurrence of second breakdown which can lead to destruction of the transistor. In this section the physical aspects of second breakdown are considered. Further, some examples of devices are presented that attempt to combine the advantages of both *DMOS* and bipolar transistors.

2.5.1 Second Breakdown

The term second breakdown is used to describe a number of breakdown mechanisms which involve a sudden decrease in transistor voltage and the constriction of current.

Safe Operating Area

The Safe Operating Area (*SOA*) of a typical bipolar power transistor is shown in Figure 2.5.1. Four limits can be distinguished. The first limit, marked *I*, defines the maximum allowable current. Beyond this current limit, damage can occur due to electromigration in the metallization or fusing of the bondwires. The second limit, marked *II*, defines the maximum allowable power dissipation. Beyond this limit thermal generation of charge carriers deteriorates the transistor performance. Usually *150°C* is chosen as maximum junction temperature. This is a practical limit to guarantee a good transistor lifetime. The third limit, marked *III*, defines the open-base breakdown voltage. The fourth limit, marked *IV*, defines the second breakdown limit. This limit is located in the high voltage low current region of operation.

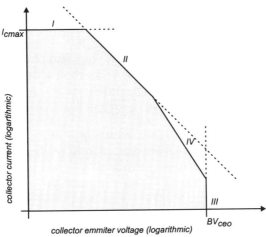

Figure 2.5.1 Safe Operating Area of a bipolar transistor

A typical property of second breakdown is that there is a time delay in its initiation. This means that the second breakdown limit can be exceeded without problems for a sufficiently short period of time. This property suggest the involvement of thermal mechanisms in second breakdown [87].

Thermal Mechanisms

A possible mechanism that can lead to second breakdown is *thermal lateral instability*. This instability results from the increase in emitter injection with temperature while a constant base-emitter voltage is applied. A local increase in current density can cause a local rise in temperature. Due to the temperature rise the local current density increases further. This regenerative situation results in constriction of the current and creation of a so-called *hot spot*. This latter breakdown mechanism is more dominant in larger transistors.

Second breakdown can also be initiated when some part of the collector-base junction exceeds the intrinsic temperature. If this occurs then the junction is short circuited.

Electrical Mechanisms

Electrical mechanism can also be responsible for second breakdown. The pinch-in effect occurs if potential gradients along the base-emitter junction lead to a regenerative situation similar to lateral thermal instability.

Avalanche injection occurs if mobile carriers form a spacecharge region which depends on the current density. In the resulting high field region avalanche generation of carriers increases the current density. Again, a regenerative situation appears.

In general, more than one mechanism can lead to second breakdown simultaneously. The importance of each mechanism depends on the transistor structure and operation mode [88-90].

Second Breakdown in DMOS Transistors

In DMOS transistors a breakdown mechanism other than the avalanche breakdown of the drain-source diode can occur. At high drain current and voltage, impact ionization at the body-drain junction causes the generation of a hole current which flows to ground through the body-region as shown Figure 2.5.2(a).

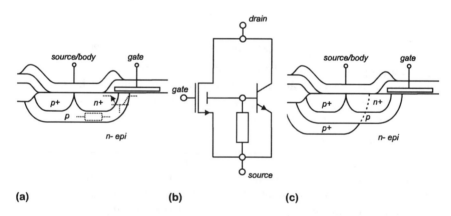

Figure 2.5.2 Second breakdown in DMOS (a) transistor structure (b) equivalent circuit (c) improved structure

Since the body-region has a nonzero resistance, a voltage drop results that causes the internal body potential to rise and, if it becomes high enough, turns on the parasitic bipolar transistor. An equivalent circuit is shown in Figure 2.5.2(b). This breakdown mechanism is sometimes also-called second breakdown although it is completely different from second breakdown in bipolar transistors.

As mentioned in Section 2.3, turn on of the parasitic bipolar transistor can also occur during fast turn-off transients of the DMOS. In both cases the body-region resistance is involved. In order to reduce the value of this resistance many *DMOS* structures have an additional *p-diffusion* in the body-region as shown in Figure 2.5.2(c). An extra feature of this diffusion is that a better control of the vertical breakdown voltage can be achieved [91-94].

2.5.2 Bipolar/DMOS Hybrids

The most important drawbacks of *DMOS* transistors compared to bipolar transistors are the lower transconductance and higher on-resistance per unit area. Some attempts have been made to combine advantageous features of both *DMOS* and bipolar in so-called *hybrid devices*.

Hybrid Conduction Mode

The parasitic bipolar can also be put to use as a functional transistor. This can be done by disconnecting the short circuit between the source/emitter and body/base and connecting the gate to the body/base as shown in Figure 2.5.3.

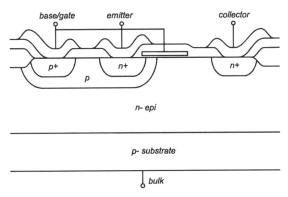

Figure 2.5.3 Hybrid conduction mode

The influence of the gate causes a reduction of the emitter-base injection barrier. This results in a hybrid conduction mode which has a higher current gain than the normal bipolar transistor [95,96].

Insulated Gate Transistor

The structure of an insulated gate is very similar to that of *DMOS* as shown in Figure 2.5.4(a). The difference between both devices is that the *n+* drain region has been replaced by a *p+* anode. If the device is turned on, the cathode-epi junction becomes forward biased and injects a hole current into the drift region that increases conductivity.

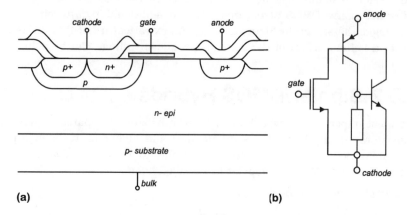

Figure 2.5.4 Insulated Gate Transistor (a) cross section (b) equivalent circuit

The parasitic bipolar transistor inherent to DMOS is turned into a parasitic thyristor in insulated gate transistors which means the device is susceptible to latch-up. An equivalent circuit is shown in Figure 2.5.4(b) [97,98].

Insulated Base Transistor

The structure of an insulated base transistor is shown in Figure 2.5.5(a). Similar to the hybrid conduction mode, the parasitic bipolar of DMOS is purposely activated in order to increase performance. An additional emitter diffusion is included in the body-region.

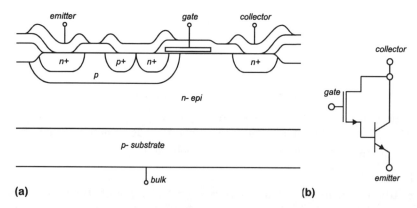

Figure 2.5.5 Insulated Base Transistor (a) cross section (b) equivalent circuit

An equivalent circuit is shown in Figure 2.5.5(b). The bipolar and DMOS transistor are connected in a Darlington configuration in which the base current of the bipolar transistor is conducted by the DMOS channel [99].

All Bipolar/DMOS hybrids have in common that they suffer from minority carrier storage and thermal instability.

2.6 Conclusion

DMOS transistors are an attractive alternative to bipolar transistors in many power applications due to their excellent thermal stability and their potential for high speed applications.

DMOS transistors have a lower transconductance and higher on-resistance than bipolar transistors with the same area. The more linear transfer characteristic of DMOS can be an advantage compared to the exponential transfer characteristic of bipolar. The more linear DMOS transistors are expected to cause inherently less distortion.

The availability of only n-channel DMOS transistors puts some special requirements on the design of output stages. In order to achieve rail-to-rail output operation the gate of the high-side DMOS has to be driven with a voltage in excess of the supply voltage.

The inherent parasitic diode and bipolar transistor have to be considered carefully when designing with DMOS. Unexpected activation of these parasitics can lead to destruction of the DMOS transistors.

2.7 References

[1] Adler, M.S., K.W. Owyang, B.J. Baliga, R.A. Kokosa, "The Evolution of Power Device Technology", *IEEE Transactions on Electron Devices*, Vol.31, No.11, pp.1570-1591, Nov. 1984

[2] Baliga, B.J., "Power Integrated Circuit - A Brief Overview", *IEEE Transactions on Electron Devices*, Vol.33, No.12, pp.1936-1939, Dec. 1986

[3] Baliga, B.J., "An Overview of Smart Power Technology", *IEEE Transactions on Electron Devices*, Vol.38, No.7, pp.1568-1575, Jul. 1991

[4] Muller, R.S., T.I. Kamins, *Device Electronics for Integrated Circuits*, 2nd Edition, John Wiley and Sons, New York, 1986

[5] Sanchez, J.J., K.K. Hsueh, T.A. DeMassa, "Drain-Engineered Hot-Electron-Resistant Device Structures: A Review", *IEEE Transactions on Electron Devices*, Vol.36, No.6, pp.1125-1132, Jun. 1989

[6] Zhou, M.-J., "Simulation of DMOSFETs", Ph.D. Dissertation, Universiteit Gent, Belgium, 1992

[7] Matsushita, T., T. Mihara, H. Ikeda, M. Hirota, Y. Hirota, "A Surge-Free Intelligent Power Device Specific to Automotive High-side Switches", *IEEE Transactions on Electron Devices*, Vol.38, No.7, pp.1576-1581, Jul. 1991

[8] Williams, R.K., A. Chang, M.E. Cornell, B. Concklin, "Design and Operation of a Fully Integrated BiC/DMOS Head-Actuator PIC for Computer Hard-Disk Drives", *IEEE Transactions on Electron Devices*, Vol.38, No.7, pp.1590-1599, Jul. 1991

[9] Zitta, H., "Smart Power Circuits for Power Switches Including Diagnostic Functions", *Proceedings AACD*, Mar 1994

[10] Fronen, R.J., J.A.M. Plagge, "A HV Control IC for Small Motors", *Proceedings ESSCIRC'93* pp.274-277, 1993

[11] Storti, S., F. Consigliere, M. Paparo, "A 30-A 30-V DMOS Motor Controller and Driver", *IEEE Journal of Solid-State Circuits*, Vol.23, No.6, pp.1394-1401, Dec. 1988

[12] Schoofs, F.A.C.M., J.C. Halberstadt, "A High-Voltage IC for a Battery Charger", *Proceedings ESSCIRC'93* pp.278-281, 1993

[13] Schoofs, F.C.A.M., "High Voltage ICs for Mains Applications", *Proceedings AACD*, Mar 1994

[14] Brasca, G., E. Botti, "A 100V/100W Monolithic Power Audio Amplifier in mixed Bipolar-MOS Technology", *IEEE Transactions on Consumer Electronics*, Vol.38, No.3, pp.217-222, Aug. 1992

[15] Blanken, P.G., P. van der Zee, "An Integrated 8MHz Video Output Amplifier", *IEEE Transactions on Consumer Electronics*, Vol.31, No.3, pp.109-118, Aug. 1985

[16] Williams, R.K., L.T. Sevilla, E. Ruetz, J.D. Plummer, "A DI/JI-Compatible Monolithic High-Voltage Multiplexer", *IEEE Transactions on Electron Devices*, Vol.33, No.12, pp.1977-1984, Dec. 1986

[17] Schoofs, F.A.C.M., C.N.G. Dupont, "A 700-V Interface IC for Power Bridge Circuits", *IEEE Journal of Solid-State Circuits*, Vol.25, No.3, pp.677-683, 1990

[18] Castello, R., F. Lari, M. Siligoni, L. Tomasini, "100V High-Performance Amplifiers in BCD Technology for SLIC Applications", *IEEE Journal of Solid-State Circuits*, Vol.27, No.9, pp.1255-1263, Sep. 1992

[19] Gariboldi, R., F. Pulvirenti, "A Monolithic Quad Line Driver for Industrial Applications", *Proceedings ESSCIRC'93* pp.282-285, 1993

[20] Sun, S.C., J.D. Plummer, "Modeling of the On-Resistance of LDMOS, VDMOS and VMOS Power Transistors", *IEEE Transactions on Electron Devices*, Vol.27, No.2, pp.356-367, Feb. 1980

[21] Stupp, E.H., S. Colak, J. Ni, Low Specific On-Resistance 400V LDMOST", *IEDM Technical Digest*, pp.426-428, 1981

[22] Claessen, H.R., P. van der Zee, "An Accurate DC Model for High-Voltage Lateral DMOS Transistors Suited for CACD", *IEEE Transactions on Electron Devices*, Vol.33, No.12, pp.1964-1970, Dec. 1986

[23] Kim, Y., J.G. Fossum, "Physical DMOST Modeling for High-Voltage IC CAD", *IEEE Transactions on Electron Devices*, Vol.37, No.3, pp.797-803, Mar. 1990

[24] Kim, Y., J.G. Fossum, R.K. Williams, "New Physical Insights and Models for High-Voltage LDMOST IC CAD", *IEEE Transactions on Electron Devices*, Vol.38, No.7, pp.1641-1649, Jul. 1991

[25] Ludikhuize, A.W., "A Versatile 250/300-V IC Process for Analog and Switching Applications", *IEEE Transactions on Electron Devices*, Vol.33, No.12, pp.2008-2015, Dec. 1986

[26] Parpia, Z., C.A.T. Samala, "Optimization of RESURF LDMOS Transistors: An Analytical Approach", *IEEE Transactions on Electron Devices*, Vol.37, No.3, pp.789-796, Mar. 1990

[27] Ludikhuize, A.W., "A Versatile 700-1200-V IC Process for Analog and Switching Applications", *IEEE Transactions on Electron Devices*, Vol.38, No.7, pp.1582-1589, Jul. 1991

[28] Appels, J.A., H.M.J. Vaes, "High Voltage Thin Layer Devices (RESURF Devices)", *IEDM Technical Digest*, pp.238-241, 1979

[29] Appels, J.A., M.G. Collet, P.A.H. Hart, H.M.J. Vaes, J.F.C.M. Verhoeven, "Thin Layer High-Voltage Devices (Resurf Devices)", *Philips Journal of Research*, Vol.35, No.1, pp.1-13, 1980

[30] Vaes, H.M.J., J.A. Appels, "High Voltage, High Current Lateral Devices", *IEDM Technical Digest*, pp.87-90, 1980

[31] Pocha, M.D., R.W. Dutton, "A Computer-Aided Design Model for High-Voltage Double-Diffused MOS (DMOS) Transistors", *IEEE Journal of Solid-State Circuits*, Vol.11, No.5, pp.718-726, Oct. 1976

[32] Colak, S., B. Singer, E. Stupp, "Lateral DMOS Power Transistor Design", *IEEE Electron Device Letters*, Vol.1, No.4, pp.51-53, Apr. 1980

[33] Colak, S., "Effects of Drift Region Parameters on the Static Properties of Power LDMOST", *IEEE Transactions on Electron Devices*, Vol.28, No.12, pp.1455-1466, Dec. 1981

[34] Mena, J.G., C.A.T. Salama, "High-Voltage Multiple-Resistivity Drift-Region LDMOS", *Solid-State Electronics*, Vol.29, No.6, pp.647-656, 1986

[35] Declercq, M.J., J.D. Plummer, "Avalanche Breakdown in High-Voltage DMOS Devices", *IEEE Transactions on Electron Devices*, Vol.23, No.1, pp.1-4, Jan. 1976

[36] Fichtner, W., J.A. Cooper, A.R. Tretola, D. Kahng, "A Novel Buried-Drain DMOSFET Structure", *IEEE Transactions on Electron Devices*, Vol.29, No.11, pp.1785-1791, Nov. 1982

References

[37] Nezar, A., C.A.T. Salama, "Breakdown Voltage in *LDMOS* Transistors Using Internal Field Rings", *IEEE Transactions on Electron Devices*, Vol.38, No.7, pp.1676-1680, Jul. 1991

[38] Stiftinger, M., W. Soppa, S. Selberherr, "A Scalable Physically Based Analytical *DMOS* Transistor Model", *Proceedings ESSDERC'94*, pp.825-828, 1994

[39] Wieder, A.W., C. Werner, J. Tihanyi, "2-D Analysis of the Negative Resistance Region of Vertical Power *MOS* Transistors", *IEDM Technical Digest*, pp.95-99, 1980

[40] Sanchez, J.L., M. Gharbi, H. Tranduc, P. Rossel, "Quasisaturation effect in high-voltage *VDMOS* transistors", *IEE Proceedings*, Vol.132, Pt.1, No.1, pp.42-46, Feb. 1985

[41] Darwish, M.N., "Study of the Quasi-Saturation Effect in *VDMOS* Transistors", *IEEE Transactions on Electron Devices*, Vol.33, No.11, pp.1710-1716, Nov. 1986

[42] Lou, K.H., C.M. Liu, J.B. Kuo, "An Analytical Quasi-Saturation Model for Vertical *DMOS* Power Transistors", *IEEE Transactions on Electron Devices*, Vol.40, No.3, pp.676-679, Mar. 1993

[43] Lou, K.H., C.M. Liu, J.B. Kuo, "77 K Versus 300 K Operation: The Quasi-Saturation Behavior of a *DMOS* Device and Its Fully Analytical Model", *IEEE Transactions on Electron Devices*, Vol.40, No.9, pp.1636-1644, Sep. 1993

[44] Liu, C.M., J.B. Kuo, Y.P. Wu, "An Analytical Quasi-Saturation Model Considering Heat Flow for a *DMOS* Device", *IEEE Transactions on Electron Devices*, Vol.41, No.6, pp.952-958, Jun, 1994.

[45] Coen, R.W., D.W. Tsang, K.P. Lisiak, "A High-Performance Planar Power *MOSFET*", *IEEE Transactions on Electron Devices*, Vol.27, No.2, pp.340-343, Feb. 1980

[46] Byrne, D.J., K. Board, "Minimization of On-Resistance of *VDMOS* Power FETs", *Electronics Letters*, Vol.19, No.14, pp.519-521, Jul. 1983

[47] Board, K., D.J. Byrne, M.S. Towers, "The Optimization of On-Resistance in Vertical *DMOS* Power Devices with Linear and Hexagonal Surface Geometries", *IEEE Transactions on Electron Devices*, Vol.31, No.1, pp.75-80, Jan. 1984

[48] Hu, C., M. Chi, V.M. Patel, "Optimum Design of Power *MOSFET*'s", *IEEE Transactions on Electron Devices*, Vol.31, No.12, pp.1693-1700, Dec. 1984

[49] Shenai, K., C.S. Korman, B.J. Baliga, P.A. Piacente, "A 50-V, 0.7-mW* cm^2, Vertical-Power *DMOSFET*", *IEEE Electron Device Letters*, Vol.10, No.3, pp.101-103, Mar. 1989

[50] Shenai, K., "Optimally Scaled Low-Voltage Vertical Power *MOSFET*s for High-Frequency Power Conversion", *IEEE Transactions on Electron Devices*, Vol.37, No.4, pp.1141-1153, Apr. 1990

[51] Stengl, J.P., H. Strack, J. Tihanyi, "Power *MOS* Transistors for 1000V Blocking Voltage", *IEDM Technical Digest*, pp.422-425, 1981

[52] Chen, X., C. Hu, "Optimum Doping Profile of Power *MOSFET* Epitaxial Layer", *IEEE Transactions on Electron Devices*, Vol.29, No.6, pp.985-987, Jun. 1982

[53] Temple, V.A.K., "Ideal FET Doping Profile", *IEEE Transactions on Electron Devices*, Vol.30, No.6, pp.619-626, Jun. 1983

[54] Darwish, M.N., K. Board, "Optimization of Breakdown Voltage and On-Resistance of *VDMOS* Transistors", *IEEE Transactions on Electron Devices*, Vol.31, No.12, pp.1769-1773, Dec. 1984

[55] Hower, P.L., M.J. Geisler, "Comparison of Various Source-Gate Geometries for Power *MOSFET*'s", *IEEE Transactions on Electron Devices*, Vol.28, No.9, pp.1098-1101, Sep. 1981

[56] Bean, K.E., "Anisotropic Etching of Silicon", *IEEE Transactions on Electron Devices*, Vol.25, No.10, pp.1185-1193, Oct. 1978

[57] Salama, C.A.T., J.G. Oakes, "Nonplanar Power Field-Effect Transistors", *IEEE Transactions on Electron Devices*, Vol.25, No.10, pp.1222-1228, Oct. 1978

[58] Tamer, A.A., K. Rauch, J.L. Moll, "Numerical Comparison of *DMOS*, *VMOS*, and *UMOS* Power Transistors", *IEEE Transactions on Electron Devices*, Vol.30, No.1, pp.73-76, Jan. 1983

[59] D'Avanzo, D.C., S.R. Combs, R.W. Dutton, "Effects of the Diffused Impurity Profile on the DC Characteristics of *VMOS* and *DMOS* Devices", *IEEE Journal of Solid-State Circuits*, Vol.12, No.4, pp.356-362, Aug. 1977

[60] Krishna, S., J. Kuo, I.S. Gaeta, "An Analog Technology Integrates Bipolar, *CMOS*, and High-Voltage *DMOS* Transistors", *IEEE Transactions on Electron Devices*, Vol.31, No.1, pp.89-95, Jan. 1984

DMOS Technology

[61] Tarng, M.L., "On-Resistance Characterization of VDMOS Power Transistors", IEDM Technical Digest, pp.429-433, 1981

[62] Wang, C.T., D.H. Navon, "Threshold and Punchthrough Behavior of Laterally Nonuniformally Doped Short-Channel MOSFET's", IEEE Transactions on Electron Devices, Vol.30, No.7, pp.776-781, Jul. 1983

[63] Pocha, M.D., J.D. Plummer, J.D. Meindl, "Tradeoff Between Threshold Voltage and Breakdown in High-Voltage Double-Diffused MOS Transistors", IEEE Transactions on Electron Devices, Vol.25, No.11, pp.1325-1327, Nov. 1978

[64] Sin, J.K.O., C.A.T. Salama, "High Frequency Distortion Analysis of DMOS Transistors", Solid-State Electronics, Vol.28, No.12, pp.1223-1233, 1985

[65] Mohan, N., T.M. Undeland, W.P. Robbins, Power Electronics:Converters, Applications and Design, John Wiley and Sons, New York, 1989.

[66] Fong, E., D.C. Pitzer, R.J. Zeman, "Power DMOS for High-Frequency and Switching Applications", IEEE Transactions on Electron Devices, Vol.27, No.2, pp.322-330, Feb. 1980

[67] Hong, M.Y., "Simulation and Fabrication of Submicron Channel Length DMOS Transistors for Analog Applications", IEEE Transactions on Electron Devices, Vol.40, No.12, pp.2222-2230, Dec. 1993

[68] Love, R.P., P.V. Gray, M.S. Adler, "A Large Area Power MOSFET Designed for Low Conduction Losses", IEDM Technical Digest, pp.418-421, 1981

[69] Chi, M-H., C. Hu, "The Operation of Power MOSFET in Reverse Mode", IEEE Transactions on Electron Devices, Vol.30, No.12, pp.1825-1828, Dec. 1983

[70] Plummer, J.D., "Monolithic MOS High Voltage Integrated Circuits", IEDM Technical Digest, pp.70-74, 1980

[71] Castro Simas, M.I., M. Simoes Piedade, J. Costa Freire, "Experimental Characterization of Power VDMOS Transistors in Commutation and a Derived Model for Computer-Aided Design", IEEE Transactions on Power Electronics, Vol.4, No.3, pp.371-378, Jul. 1989

[72] Shimada, Y., K. Kato, S. Ikeda, H. Yoshida, "Low Input Capacitance and Low Loss VD-MOSFET Rectifier Element", IEEE Transactions on Electron Devices, Vol.29, No.8, pp.1332-1334, Aug. 1982

[73] Ueda, D., H. Takagi, G. Kano, "A New Vertical Double Diffused MOSFET - The Self-Aligned Terraced-Gate MOSFET", IEEE Transactions on Electron Devices, Vol.31, No.4, pp.416-420, Apr. 1984

[74] Sakai, T., N. Murakami, "A New VDMOSFET Structure with Reduced Reverse Transfer Capacitance", IEEE Transactions on Electron Devices, Vol.36, No.7, pp.1381-1386, Jul. 1989

[75] Lu, C., N. Tsai, C.N. Dunn, P.C. Riffe, M.A. Shibib, R.A. Furnanage, C.A. Goodwin, "An Analog/Digital BCDMOS Technology with Dielectric Isolation - Devices and Processes", IEEE Transactions on Electron Devices, Vol.35, No.2, pp.230-239, Feb. 1988

[76] Apel, U., H.G. Graf, C. Harendt, B. Hofflinger, T. Ifstrom, "A 100-V Lateral DMOS Transistor with a 0.3-Micrometer Channel in a 1-Micrometer Silicon-Film-on-Insulator-on-Silicon", IEEE Transactions on Electron Devices, Vol.38, No.7, pp.1655-1659, Jul. 1991

[77] Ifstrom, T., U. Apel, H. Graf, C. Harendt, B. Hofflinger, "A 150-V Multiple Up-Drain VDMOS, CMOS and Bipolar Process in "Direct-Bonded" Silicon on Insulator on Silicon", IEEE Electron Device Letters, Vol.13, No.9, pp.460-461, Sep. 1992

[78] Baliga, B.J., "High-Voltage Device Termination Techniques; A Comparative Review", IEE Proceedings, Vol.129, Pt.I, No.5, pp.173-179, Oct. 1982

[79] Lane, W.A., C.A.T. Salama, "Epitaxial VVMOS Power Transistors", IEEE Transactions on Electron Devices, Vol.27, No.2, pp.349-355, Feb. 1980

[80] Korec, J., R. Held, "Comparison of DMOS/IGBT-Compatible High-Voltage Termination Structures and Passivation Techniques", IEEE Transactions on Electron Devices, Vol.40, No.10, pp.1845-1854, Oct. 1993

[81] Tantraporn, W., V.A.K. Temple, "Multiple-Zone Single-Mask Junction Termination Extension - A High-Yield Near-Ideal Breakdown Voltage Technology", IEEE Transactions on Electron Devices, Vol.34, No.10, pp.2200-2210, Oct. 1987

[82] Yamaguchi, T., S. Morimoto, "Process and Device Design of a 1000-V MOS IC", IEEE Transactions on Electron Devices, Vol.29, No.8, pp.1171-1178, Aug. 1982

References

[83] Andreini, A., C. Contiero, P. Galbiati, "A New Integrated Silicon Gate Technology Combining Bipolar Linear, CMOS Logic and DMOS Power Parts", *IEEE Transactions on Electron Devices*, Vol.33, No.12, pp.2025-2030, Dec. 1986

[84] Chang, M.F., G. Pifer, H. Yilmaz, E.J. Wildi, R.G. Hodgins, K. Owyang, M.S. Adler, "Lateral HVIC with 1200-V Bipolar and Field-Effect Devices", *IEEE Transactions on Electron Devices*, Vol.33, No.12, pp.1992-2001, Dec. 1986

[85] Elfand, T., T. Keller, S. Keller, J. Rodriguez, "Optimized Complementary $0V Power LDMOSFETs Use Existing Fabrication Steps In Submicorn CMOS Technology", *IEDM Technical Digest*, pp.399-402, 1994

[86] Thomas, G., G. Troussel, F. Vialettes, "High-Voltage Technology Offers New Solutions for Interface Integrated Circuits", *IEEE Transactions on Electron Devices*, Vol.33, No.12, pp.2016-2024, Dec. 1986

[87] Murari, B., "Power Integrated Circuits: Problems, Tradeoffs, and Solutions", *IEEE Journal of Solid-State Circuits*, Vol.13, No.3, pp.307-319, Jun. 1978

[88] Schafft, H.A., "Second Breakdown - A Comprehensive Review", *Proceedings of the IEEE*, Vol.55, No.8, pp.1272-1288, Aug. 1967

[89] Hower, P.L., V.G.K. Reddi, "Avalanche Injection and Second Breakdown in Transistors", *IEEE Transactions on Electron Devices*, Vol.17, No.4, pp.320-335, Apr. 1970

[90] Hwang, K., D.H. Navon, T.-W. Tang, P.L. Hower, "Second Breakdown Prediction by Two-Dimensional Numerical Analysis of BJT Turnoff", *IEEE Transactions on Electron Devices*, Vol.33, No.7, pp.1067-1072, Jul. 1986

[91] Krishna, S., "Second Breakdown in High Voltage MOS Transistors", *Solid-State Electronics*, Vol.20, pp.875-878, 1977

[92] Hu, C., M. Chi, "Second Breakdown of Vertical Power MOSFET's", *IEEE Transactions on Electron Devices*, Vol.29, No.8, pp.1287-1293, Aug. 1982

[93] Yoshida, I., T. Okabe, M. Katsueda, S. Ochi, M. Nagata, "Thermal Stability and Secondary Breakdown in Planar Power MOSFET's", *IEEE Transactions on Electron Devices*, Vol.27, No.2, pp.395-398, Feb. 1980

[94] Amerasekera, A., L. van Rozendaal, J. Bruines, F. Kuper, "Characterization and Modeling of Second Breakdown in NMOST's for the Extraction of ESD-Related Process and Design Parameters", *IEEE Transactions on Electron Devices*, Vol.38, No.9, pp.2161-2168, Sep. 1991

[95] Edholm, B., J. Olsson, A. Söderbärg, K. Bohlin, F. Magnusson, "High Current Gain Lateral Bipolar Action in DMOS Transistors", *Proceedings ESSDERC'94*, pp.221-224, 1994

[96] Fischer, K.J., K. Shenai, "Effect of Bipolar Turn-On on the Static Current-Voltage Characteristics of Scaled Power DMOSFET's", *IEEE Transactions on Electron Devices*, Vol.42, No.3, pp.555-563, Mar. 1995

[97] Leipold, L., W. Baumgartner, W. Ladenhauf, J.P. Stengl, "A FET Controlled Thyristor in SIPMOS Technology", *IEDM Technical Digest*, pp.79-82, 1980

[98] Tihanyi, J., "Functional Integration of Power MOS and Bipolar Devices", *IEDM Technical Digest*, pp.75-78, 1980

[99] Parpia, Z., J.G. Mena, C.A.T. Salama, "A Novel CMOS-Compatible High-Voltage Transistor Structure", *IEEE Transactions on Electron Devices*, Vol.33, No.12, pp.1948-1952, Dec. 1986

3

Chargepump Circuits

This chapter deals with voltage multipliers. These are circuits that can generate a voltage higher than the supply voltage without the use of inductors. Some well-known voltage multiplication techniques are presented and compared on their suitability for integration. Based on this comparison a specific voltage multiplier circuit called a *chargepump* is selected. A technique is presented to reduce the output voltage ripple of chargepumps. Some methods are discussed for regulation of the output voltage level. A fully integrated chargepump design is presented and discussed.

3.1 Introduction

Most electronic circuits require a *DC* power source. A *DC* power source provides a predetermined *DC* voltage which is as independent as possible of the current drawn from the source. The value of the *DC* supply voltage is often not optimal for the electronic circuit it is supplying power to. In some cases, more than one *DC* voltage is required for optimal power efficiency. To make these *DC* voltages, the efficient conversion of one *DC* voltage to another is useful. Conversion of a *DC* voltage to a higher *DC* voltage is called *step-up conversion* or *boost conversion*.

3.1.1 Switching Voltage Regulators

A widely used, and very efficient step-up converter is the *switching voltage regulator* shown in Figure 3.1.1. The operation of this converter can be understood as follows. During the time interval T_1 the switch S is closed. The diode D is reverse-biased by the positive output voltage V_Q and the capacitor C discharges through the load R_L. By choosing $CR_L >> T_1$, the drop in V_Q (the ripple voltage) is small. During this interval the supply voltage V_P is across the inductor L and the current i_L through the inductor increases by $di_L = dtV_P/L = T_1 V_P/L$.

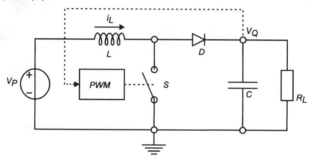

Figure 3.1.1 Switching voltage regulator.

Next, consider the time interval T_2 during which the switch S is open. Since the current in an inductor cannot change instantaneously, the diode D becomes forward-biased, and i_L passes through the diode D and into capacitor C. In this way the output voltage V_Q is raised. In steady-state the voltage across C must be the same at the end of the period $T=T_1+T_2$ as it was in the beginning. The value of the steady-state output voltage V_Q is regulated using pulse-width-modulation (*PWM*) and feedback. An efficiency of more than *90%* can easily by achieved with this type of step-up converter. However, it has two important drawbacks. First, unlike the diode, switch and capacitor, the inductor cannot be integrated [1]. Second, the high *dV/dt's* that occur during switching can cause *EM-radiation*.

3.1.2 Voltage Multiplication

DC-DC step-up converters that do not use inductors are based on *bootstrapping*. The basic idea of bootstrapping is demonstrated in Figure 3.1.2. First a capacitor C is charged with the supply voltage V_P. Then the capacitor is disconnected from the voltage source. Finally, the bottom plate of the capacitor is connected to the supply voltage V_P. Now the top plate voltage V_Q equals $2V_P$. The capacitor C seems to be "pulling up itself by its bootstraps". This bootstrap can be repeated to generate higher voltages.

Introduction

The class of circuits based on this technique is called *voltage multipliers* since the unloaded output voltage V_Q is always an integer multiple of the supply voltage V_P. The output voltage V_Q drops if a current I_o has to be delivered to a load. A large storage capacitor C_o is usually connected to the output of the voltage multiplier to ensure a sufficient constant DC output voltage.

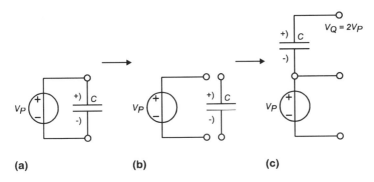

Figure 3.1.2 Bootstrapping.

The operation of a voltage multiplier can be described by two alternating phases. In the first phase, the *charge phase*, the capacitors in the network are connected in such a way that they are charged by the supply voltage V_P. During this phase, the load current I_o is provided by the storage capacitor C_o. In the second phase, the *bootstrap phase*, the capacitors are reconnected in such a way that the desired output voltage is generated. During this phase the voltage multiplier provides the load current I_o and recharges the storage capacitor C_o. When these two phases are alternated with a frequency f, a constant output current can be sourced.

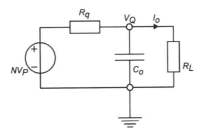

Figure 3.1.3 Equivalent voltage multiplier circuit.

As stated earlier, the unloaded output voltage V_Q is an integer multiple of the supply voltage V_P. However, this voltage can only be reached when the voltage multiplier is not loaded. A load current causes the output voltage

Chargepump Circuits

V_Q to drop due to charge redistribution losses in the voltage multiplier network. These losses can be modeled with an equivalent output resistance R_q. An equivalent circuit of a voltage multiplier is shown in Figure 3.1.3.

Voltage multipliers can be realized using only capacitors, diodes and/or switches so they are well suited for integration. Due to the pumplike operation of these circuits they are often called *chargepumps*.

3.1.3 Application Areas of Chargepumps

Integrated chargepumps can be used when a voltage higher than the supply voltage is needed. This second supply voltage may be too expensive so the desired voltage has to be generated on-chip from the available supply voltage. Two typical application areas of on-chip chargepumps are Electrically Erasable and Programmable Read-Only Memories (*EEPROM*) and *NMOS* power switches.

EEPROM

The basic principle of solid-state nonvolatile memories is that the logic state of the memory cell is determined by the threshold voltage of a *NMOS* transistor. The threshold voltage can be varied by changing the amount of charge that is stored in the gate oxide [2]. Two device types can be used to achieve this. The first type is a Metal Nitride Oxide Silicon (*MNOS*) transistor. The gate oxide of this transistor consists of a very thin silicondioxide layer on which a silicon nitride layer is deposited. At the oxide/nitride interface electrons can be trapped.

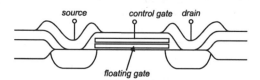

Figure 3.1.4 EEPROM floating gate structure.

The second, more popular type is a so-called floating gate transistor as shown in Figure 3.1.4. The charge stored on the floating gate determines the effective threshold voltage of the device. Most *EEPROM*s are based on floating gate devices.

To charge and discharge the oxide/nitride interface or floating gate, charge carriers have to be transported through the insulating oxide layer. Two different mechanisms are used in practical devices: Fowler-Nordheim tunneling and hot-electron injection.

For both mechanisms a high electric field is required. In order to realize this high electric field, a voltage in the range of *15V* to *20V* is needed while the supply voltage is typically in the range of *3V* to *5V*. This high voltage is usually generated by a chargepump [2-14]. Chargepumps used in *EEPROM*s typically need a large number of stages (up to *26*) in order to generate a multiple of the supply voltage.

NMOS Power Switches

Power switches can be realized with Vertical Double-diffused *MOS* (*VDMOS*) transistors. *VDMOS* transistors have fast switching capability, high breakdown voltage and excellent thermal stability. These features make them very suitable for power applications. In order to turn a *DMOS* transistor maximally on, a gate-source voltage is needed that is higher (*~10V*) than the supply voltage. When *DMOS* transistors are used in fast switching applications, such as pulse-width-modulation (*PWM*), a bootstrap capacitor is often used to provide this gate voltage. However, for *DC* operation this is not a practical solution. In this case, the gate voltage can be generated by a chargepump [15-18]. Chargepumps in this application area usually have to be able to operate over a wide supply voltage range.

Low Voltage Clock Drive Boosting

A problem that is similar to the *NMOS* power switches occurs in low voltage applications. The application of *MOS* transistors as analog switches or transmission gates has proven to be very efficient for implementing all kinds of analog functions. The conduction of a *MOS* transistor depends, among other things, on the gate voltage that is applied. Due to reduction of the supply voltage a nonconduction gap is created in the mid-range of the supply voltage. This problem can be overcome by boosting the clock voltage for the *NMOS* transistor or for both *NMOS* and *PMOS* transistors [19-23]. In general, the dynamic range of an analog circuit can be increased by lifting the gate voltages of *MOS* transistors above the supply voltage [24].

3.2 Voltage Multipliers

The first voltage multiplier arrangements were designed by physicists for use in the laboratory for charged particle acceleration or spark generation. These voltage multipliers were used to generate voltages higher than those which could be easily handled by electromagnetic transformers. This is possible since the maximum voltage across any of the capacitors is only equal to the supply voltage, irrespective of the number of multiplying stages, as is explained later. Later voltage multipliers also became popular in *TV* applications to generate high voltages for driving cathode-ray tubes.

Integration of voltage multipliers became important with the advent of nonvolatile memories.

A voltage multiplier network consists of capacitors and switches and/or diodes. A lot of alternative voltage multiplier network topologies have been developed. Generation of these topologies can be done systematically [25,26]. To gain some insight into the various topological possibilities a short description of three well-known voltage multiplier arrangements is presented and compared in the following.

3.2.1 Marx Voltage Multiplier

The Marx voltage multiplier is based on the Marx high voltage generator that is used in laboratory experiments to generate unipolar pulses of very high voltage (up to several million volts). In this high voltage impulse generator, a bank of capacitors is charged in parallel, via resistors, from a *DC* supply, and discharged in series via a string of spark gaps. The same principle can be used in order to make a *DC-DC* voltage multiplier by replacing the resistors and spark gaps by switches and adding a storage capacitor to the output [27-29].

The Marx voltage multiplier is based on parallel-series switching of capacitors. First, in the *charge phase*, *N-1* capacitors are charged in parallel with supply voltage V_P. Then, in the *bootstrap phase*, these *N-1* capacitors are connected in series with the supply voltage V_P, resulting in an output voltage $V_Q=NV_P$. The storage capacitor C_o is loaded by the output current I_o resulting in a charge loss of I_o/f per clock cycle. In the bootstrap phase the storage capacitor C_o is recharged with charge Q. A steady-state is reached if Q equals I_o/f. A Marx voltage sextupler during the charge and bootstrap phase is shown in Figure 3.2.1. The concept of series-parallel switching can also be reversed in order to perform voltage division.

An expression for the steady-state average output voltage V_Q of the Marx voltage multiplier is easily found to be:

$$V_Q = N \cdot V_P - \frac{N-1}{C \cdot f} \cdot I_o$$

(3.2.1)

where *N* is the voltage multiplication factor, *C* is the value of the pump capacitors and *f* is the switching frequency. From this expression it can be seen that the equivalent output resistance R_q of the Marx voltage multiplier is:

$$R_q = \frac{N-1}{C \cdot f}$$

(3.2.2)

and increases linearly with the multiplication factor N.

The advantages of the Marx voltage multiplier become apparent if it is used in the original form as a pulse generator. First, the voltage across each of the pump capacitors is never higher than the supply voltage V_P. This means that an output voltage can be much higher than the breakdown voltage of the single capacitors. Second, the pump capacitors are charged in parallel so the circuit can recharge very fast. However, these advantages disappear when a storage capacitor C_o is added in order to supply a constant output voltage. This capacitor C_o has to withstand the entire output voltage V_Q.

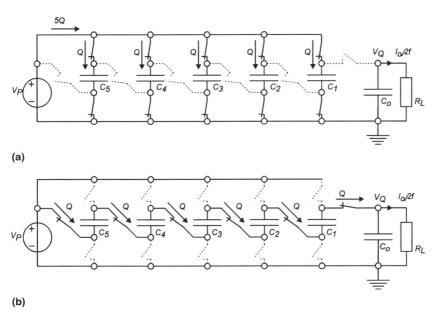

Figure 3.2.1 Marx voltage sextupler (a) charge phase (b) bootstrap phase.

A disadvantages of the Marx voltage multiplier is the complexity of switching involved. The parallel-series switching requires a large number of switches and accurate timing. This would make an implementation of a Marx voltage multiplier rather complex. This type of voltage multiplier is less suitable for monolithic integration since on-chip capacitors have relative high parasitic capacitance between the bottom plate and substrate. The charging and discharging of these parasitics can lead to substantial losses, especially towards the output of the multiplier where the voltage excursions can become quite large.

3.2.2 Cockcroft-Walton Voltage Multiplier

Like the Marx generator, the Cockcroft-Walton voltage multiplier was originally used for high voltages for laboratory experiments. Further, Cockcroft-Walton voltage multipliers have been used extensively in the high voltage power supply of television sets. In such applications, the multiplier is usually connected directly to the secondary winding of a transformer. The *AC* voltage is then rectified and multiplied to several kiloVolts. In contrast to Marx generators, Cockcroft-Walton voltage multipliers do not generate pulses but can deliver a constant *DC* voltage [30-33].

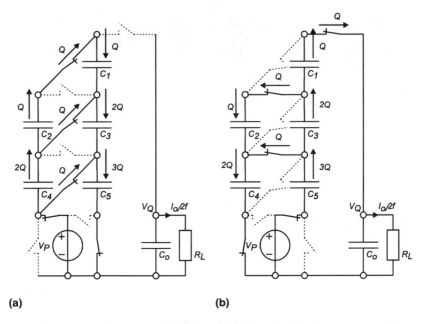

Figure 3.2.2 Cockcroft-Walton voltage sextupler (a) charge phase (b) bootstrap phase.

A Cockcroft-Walton voltage multiplier basically consists of two stacks of capacitors which are mutually interconnected by diodes or switches. All charge that reaches the output of the multiplier is pumped up from the bottom through the entire circuit. This makes a Cockcroft-Walton multiplier inherently slower in reaching steady-state than a Marx voltage multiplier. A Cockcroft-Walton voltage sextupler during the charge and bootstrap phase is shown in Figure 3.2.2.

An expression for the steady-state average output voltage V_Q of the Cockcroft-Walton voltage multiplier is given by:

$$V_Q = N \cdot V_P - \frac{\frac{N}{6} \cdot \left(\frac{N^2}{2} + 1\right)}{C \cdot f} \cdot I_o$$

(3.2.3)

where N is the voltage multiplication factor, C is the value of the pump capacitors and f is the switching frequency. So, the equivalent output resistance of the Cockcroft-Walton voltage multiplier is:

$$R_q = \frac{\frac{N}{6} \cdot \left(\frac{N^2}{2} + 1\right)}{C \cdot f}$$

(3.2.4)

and increases for high N with the cubic of the multiplication factor N. Clearly, the output resistance of the Cockcroft-Walton voltage multiplier increases much faster than that of the Marx voltage multiplier. However, in these calculations, all capacitors are assumed to be of equal value. In Figure 3.2.2 it can be seen that the currents flowing through the capacitors are not equal. If the capacitors values are scaled proportionally to their current, the change in voltage across each capacitor is the same. In this case the expression for the output voltage V_Q becomes:

$$V_Q = N \cdot V_P - \frac{N^2}{4 \cdot C \cdot f} \cdot I_o$$

(3.2.5)

and the equivalent output resistance equal to:

$$R_q = \frac{N^2}{4 \cdot C \cdot f}$$

(3.2.6)

which is a substantial improvement.

In most applications of the Cockcroft-Walton voltage multiplier, the storage capacitor C_o is omitted. In that case the voltage across each of the pump capacitors is always less than two times the supply voltage V_P which is, similar to the Marx voltage multiplier, an advantage in high voltage applications. An advantage compared to the Marx voltage multiplier is that the Cockcroft-Walton voltage multiplier can be constructed much simpler using diodes instead of switches.

Similar to the Marx voltage multiplier, the Cockcroft-Walton voltage multiplier is rather sensitive to parasitics which make it a poor candidate

Chargepump Circuits

for integration. Further, the equivalent output resistance is rather high and grows rapidly with the number of stages.

3.2.3 Dickson Voltage Multiplier

The Marx and Cockcroft-Walton voltage multipliers have in common that they are both sensitive to parasitic capacitance which is a severe problem for application in integrated circuits. A voltage multiplier that is less sensitive to parasitic capacitance is the Dickson voltage multiplier. The original Dickson voltage multiplier was designed for application in EEPROM devices. Also discrete realizations suited for higher output powers are possible [34-44].

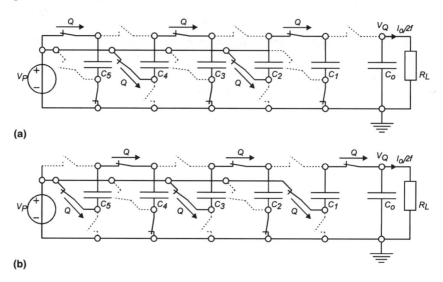

Figure 3.2.3 Dickson voltage sextupler (a) charge phase (b) bootstrap phase

The operation of the Dickson voltage multiplier is very similar to that of the Cockcroft-Walton voltage multiplier. All charge that reaches the output of the multiplier is pumped up from the bottom through the entire circuit. However, the difference is that the capacitors are not stacked but are connected in parallel. The bottom plates of the capacitors are connected to the supply voltage V_P or to ground. Consequently, the charging and discharging of the substrate parasitic is performed by the supply voltage source so no charge is lost in the voltage multiplication. A Dickson voltage sextupler during the charge and bootstrap phase is shown in Figure 3.2.3.

The steady-state average output voltage V_Q of the Dickson voltage multiplier is described by the same expression as the Marx voltage multiplier:

$$V_Q = N \cdot V_P - \frac{N-1}{C \cdot f} \cdot I_o$$

(3.2.7)

and therefore has the same low equivalent output resistance:

$$R_q = \frac{N-1}{C \cdot f}$$

(3.2.8)

The most important advantage of the Dickson voltage multiplier is that it is insensitive to parasitic capacitance. Further, like the Cockcroft-Walton voltage multiplier, it can be realized easily using diodes instead of switches. In this case no complex switching is needed. In integrated circuits the diodes are often replaced by *MOS* transistors connected as a diode or *MOS* switches which have lower on-resistance. Clearly the Dickson voltage multiplier is the most interesting candidate for integration.

The capacitors of the Dickson voltage multiplier have to withstand the increasing voltages towards the output as they are build up along the chain. This makes the Dickson voltage multiplier unattractive for high voltage applications.

3.2.4 Comparison of Voltage Multipliers

Besides the three voltage multipliers described in this section, other arrangements are also possible. However, these alternative arrangements do not have any advantage with respect to integration. All voltage multipliers have in common that a multiplication with a factor *N*, requires *N-1* multiplying capacitors. If only doubling of the supply voltage is required then all voltage multiplier reduce to the same circuit. If tripling is required then the Cockcroft-Walton and Dickson voltage multipliers are the same. The difference between different arrangements appears as the number of stages is increased. In the following sections only the Dickson voltage multiplier or *chargepump* will be considered since it is best suited for integration.

3.3 Chargepump Operation

Most integrated chargepump circuits used are based on the circuit proposed by Dickson in 1976 [34]. The basic electrical circuit of this chargepump is shown in Figure 3.3.1. The circuit consists of a chain of capacitors C_n which are interconnected by diodes D_n and coupled in parallel with two noninterleaving clock signals $\Phi(t)$ and $\overline{\Phi}(t)$ with frequency *f* and amplitude V_Φ. Charge is transferred from one capacitor to

Chargepump Circuits

the next if the clock driving the first capacitor is high, and the clock driving the next capacitor is low.

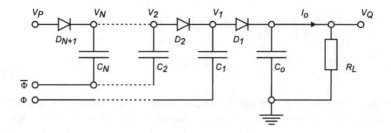

Figure 3.3.1 *Basic Dickson chargepump circuit.*

A simple expression for the steady-state output voltage V^Ω_Q is given by:

$$V^\Omega_Q = V_P - (N+1) \cdot V_\Delta + N \cdot V_\Phi - \left(\frac{N}{C} + \frac{1}{2 \cdot C_o}\right) \cdot \frac{I_o}{f}$$

(3.3.1)

where V_Δ is the forward-bias voltage of the diodes, N is the number of stages and C is the size of the pump capacitors. Note that this expression gives the minimal value of the output voltage instead of the average value. A detailed derivation of this expression is described in Chapter 4.

In the original Dickson chargepump the pump capacitors are driven with square wave voltages. The steep voltage transients cause peak currents and a ripple on the output voltage. These undesirable effects can be reduced by using a different driving regime as will be demonstrated in the following paragraphs. The analysis is confined to *1-stage* chargepump circuits, that is, chargepumps with only one pump stage. However, the improvements presented in this section are also applicable to chargepumps with more stages.

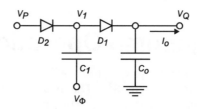

Figure 3.3.2 *1-stage chargepump circuit.*

3.3.1 Normal Mode

The basic circuit of a *1-stage* chargepump circuit is shown in Figure 3.3.2. The steady-state output voltage V^Ω_Q of a *1-stage* chargepump is given by:

$$V^\Omega_Q = V_P - 2 \cdot V_\Delta + V_\Phi - \left(\frac{1}{C_1} + \frac{1}{2 \cdot C_o}\right) \cdot \frac{I_o}{f}$$

(3.3.2)

If the circuit is loaded with a constant current I_o the ripple voltage V^ω_Q at the output due to discharging of the storage capacitor C_o has a sawtooth shape as shown in Figure 3.3.3. The amplitude of the ripple voltage V^ω_Q is given by:

$$V^\omega_Q = \left(\frac{1}{C_1 + C_o} + \frac{1}{C_o}\right) \cdot \frac{I_o}{2 \cdot f} \approx \frac{I_o}{f \cdot C_o}$$

(3.3.3)

The sharp rising edge is caused by the (almost) instantaneous charge transfer between pump capacitor C_1 and storage capacitor C_o. The (piecewise) linear voltage drop is caused by the discharging of the output capacitor C_o.

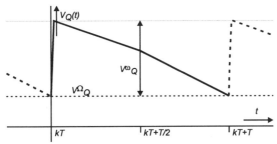

Figure 3.3.3 Output voltage waveform.

During the first half of the clock period diode D_1, which connects pump capacitor C_1 and storage capacitor C_o, conducts. Consequently, C_1 and C_o are connected in parallel resulting in a larger capacitance. In the second half clock period diode D_1 is off and only C_o is discharged. This explains the difference in dV_Q/dt between both halves of a clock period.

3.3.2 Double Phase Mode

The voltage ripple can be reduced by distributing the charge transfer towards the storage capacitor C_o over a longer period of time. This can be done by using a double phase chargepump as shown in Figure 3.3.4.

Chargepump Circuits

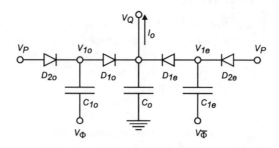

Figure 3.3.4 Double phase 1-stage chargepump circuit.

The double phase chargepump basically consists of two chargepumps connected in parallel and driven in antiphase. The original pump capacitor C_1 is split into two equal parts C_{1o} and C_{1e}:

$$C_{1o} = C_{1e} = \frac{C_1}{2}$$

(3.3.4)

The resulting steady-state output voltage V^{Ω}_Q is given by:

$$V^{\Omega}_Q = V_P - 2 \cdot V_{\Delta} + V_{\Phi} - \frac{I_o}{f \cdot C_1}$$

(3.3.5)

This steady-state value is little higher than the steady-state output voltage of the normal, single phase chargepump. In a single phase chargepump, the storage capacitor is isolated from the rest of the circuit during the second half of a clock period.

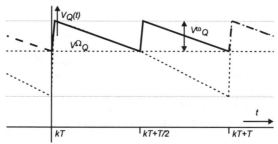

Figure 3.3.5 Output voltage waveform of a double phase chargepump.

During this period, the storage capacitor C_o is discharged by the load current I_o causing a voltage drop. In a double phase chargepump, charge is transferred to the storage capacitor during both half clock periods.

Therefore the output voltage during the second half clock period is identical to the output voltage during the first half clock period. The resulting output voltage waveform is shown in Figure 3.3.5. The dotted line indicates the output voltage waveform for a single phase chargepump. The amplitude of the output voltage ripple V^ω_Q is:

$$V^\omega_Q = \left(\frac{1}{C_o + \frac{1}{2} \cdot C_1}\right) \cdot \frac{I_o}{2 \cdot f} \approx \frac{1}{2} \cdot \frac{I_o}{f \cdot C_o}$$

(3.3.6)

So a reduction of the ripple with more than a factor 2 can be achieved with this simple modification.

3.3.3 Current Driven Mode

Another method to reduce the output voltage ripple is to prevent the charge transfer to the storage capacitor from being abrupt. A simple method to do this is to drive the pump capacitor with a constant current I_Φ instead of a square wave voltage V_Φ.

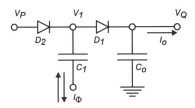

Figure 3.3.6 Current driven 1-stage chargepump circuit.

As long as the driving current I_Φ is at least twice as large as the load current I_o, the steady-state output voltage V^Ω_Q is identical to that of the original chargepump. In this case the current driver will saturate and the amplitude of the voltage at the output of the driver will be equal to the supply voltage as was the case with the normal voltage driven chargepump. For a 1-stage chargepump, the steady-state output voltage V^Ω_Q remains unchanged:

$$V^\Omega_Q = V_P - 2 \cdot V_\Delta + V_\Phi - \left(\frac{1}{C_1} + \frac{1}{2 \cdot C_o}\right) \cdot \frac{I_o}{f}$$

(3.3.7)

If the driving current I_Φ exactly equals twice the load current I_o the ripple is minimal. In this case, the current flowing towards the storage capacitor C_o

Chargepump Circuits

during the first half clock period is equal to the current flowing from the storage capacitor in the second half clock period resulting in a triangular ripple. The output voltage waveform for a number of drive currents is shown in Figure 3.3.7. The amplitude of the output voltage ripple V^ω_Q is given by:

$$\frac{1}{C_o} \cdot \frac{I_o}{2 \cdot f} \leq V^\omega_Q \leq \left(\frac{1}{C_o + C_1} + \frac{1}{C_o}\right) \cdot \frac{I_o}{2 \cdot f}$$

(3.3.8)

and depends on the drive current I_Φ.

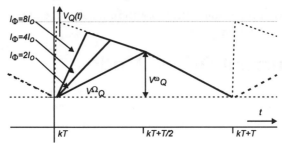

Figure 3.3.7 Output voltage waveform of a current driven chargepump

So again a reduction of a factor two can be achieved. An additional advantage of the current driven mode is that the peak currents through the diodes become much smaller. In fact, all dV/dt's in the circuit are smaller so it can be expected that the circuit will produce less noise and interference with other circuitry on the same chip.

3.3.4 Double Phase Current Driven Mode

Logically, the next step is to combine the double phase and current driven modes. In this case the charge transfer towards the storage capacitor C_o is spread out maximally in time. The corresponding circuit is shown in Figure 3.3.8.

As long as the driving current I_Φ on both sides of the chargepump are higher than or equal to the load current I_o, the dynamic behavior of this chargepump equals that of the double phase voltage driven chargepump. The steady-state output voltage V^Ω_Q of a 1-stage double phase current driven chargepump is then given by:

$$V^\Omega_Q = V_P - 2 \cdot V_\Delta + V_\Phi - \frac{I_o}{f \cdot C_1}$$

(3.3.9)

Chargepump Operation

Minimal output voltage ripple requires the driving currents I_Φ to be exactly equal to the load current I_o. In this case, the current flowing towards the storage capacitor C_o during both clock phases is zero and so is the voltage ripple.

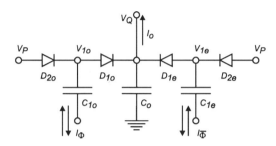

Figure 3.3.8 Double phase current driven 1-stage chargepump circuit.

The amplitude of the output voltage ripple V^ω_Q is:

$$0 \leq V^\omega_Q \leq \left(\frac{1}{C_o + \frac{1}{2} \cdot C_1}\right) \cdot \frac{I_o}{2 \cdot f}$$

(3.3.10)

and depends on the drive current I_Φ as shown in Figure 3.3.9.

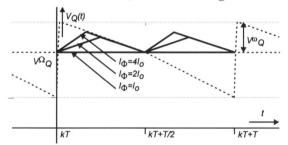

Figure 3.3.9 Output voltage waveform of a double phase current driven chargepump.

Another advantageous effect of this mode is that the current drawn from the supply voltage V_P is also constant. Of course, the ideal zero voltage ripple cannot be achieved in a practical circuit due to the presence of parasitics and mismatch. However, a substantial reduction of the voltage ripple is possible.

3.3.5 Simulation Results

In order to verify the qualitative aspects described in the previous sections, some circuit simulations have been performed. In the simulation idealized components have been used. In Figure 3.3.10 the steady-state output voltage waveforms of a *1-stage* chargepump circuit are shown for the four operation modes.

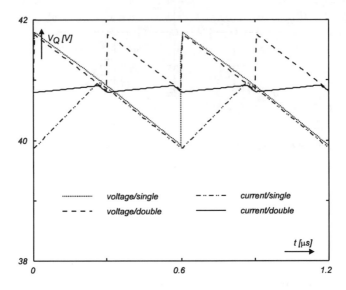

Figure 3.3.10 Output voltage waveform for different modes of operation.

The other parameters used for this simulation are: $f=1.67MHz$, $I_o=1.0mA$, $C_{1o}=C_{1e}=16.7pF$, $C_o=300pF$, $V_P=30V$ and $V_\Delta=0.6V$. For the simulation of the current driven modes, the drive current I_Φ has been matched to the load current I_o. In a practical circuit this matching will require some kind of control mechanism. From Figure 3.3.10 can be seen that the simulation agrees very well with the theory.

3.4 Voltage Control

The steady-state output voltage V^Ω_Q of a chargepump circuit depends on a number of parameters. As was mentioned in the previous section the steady-state output voltage of an *N-stage* chargepump is given by:

$$V^\Omega_Q = V_P - (N+1) \cdot V_\Delta + N \cdot V_\Phi - \left(\frac{N}{C} + \frac{1}{2 \cdot C_o} \right) \cdot \frac{I_o}{f}$$

(3.4.1)

In many applications the output voltage of a chargepump should have a well defined value that is independent of the load current I_o. This requires some kind of voltage regulation. In this section, a number of regulation techniques found in literature are discussed. From equation (3.4.1) it can be seen that in order to regulate the output voltage, at least one of the parameters V_P, V_Φ, N, f, C, or I_o has to be controlled. The value of V_Δ is determined by the technology used. Further, the value of $1/2C_o$ is usually much smaller than N/C. So changing the value of C_o is not a very effective means of regulation.

3.4.1 Output Voltage Clipping

The simplest method to regulate the output voltage of a chargepump is to clip it with a Zenerlike clipping circuit. This method is often used in *EEPROM* applications [6-9]. A rather large number (*6* to *26*) of stages is used that can provide a high nominal output voltage that has to be regulated down to prevent junction breakdown. Clearly, this voltage regulation method is not very efficient with respect to power consumption since the pump always has to deliver the maximal current possible at the regulated output voltage. The current that is not needed by the load is dissipated in the clipping circuit.

3.4.2 Amplitude Control

Clipping circuits can also be used to limit the amplitude of the supply and clock voltages V_P and V_Φ. This method is used in *DMOS* driver circuits where the required chargepump output voltage V_Q is not an integer multiple of the supply voltage but has to be about *10V* higher than the supply voltage. This voltage is necessary to turn the high-side *DMOS* transistor maximally on. This is done by using a *1-stage* or *2-stage* chargepump with limited supply voltage. The supply voltage is limited, referred to the main supply [15] or ground [17] with a Zenerlike clipping circuit. An advantage of this method is that the complete circuit can work under a wide range of main supply voltages. This method can be considered a form of clock amplitude control.

A more sophisticated form of amplitude control can be achieved by using feedback. Oto, *et al.* [5] use a feedback loop that is made up of a clock driver with controllable output swing, the chargepump circuit, a voltage divider and a differential amplifier. Negative feedback forces the output voltage to be equal to a reference voltage multiplied by the reciprocal of the divider ratio.

A form of duty-cycle modulation to control the clock voltage swing is used in [38,39]. The switch timing is controlled in such a way that the capacitors are not charged with the full supply voltage. This can be done because the charging of the pump capacitors is not instantaneous due to series resistance.

3.4.3 Frequency Control

According to Goryunov [40] the frequency of a chargepump circuit can be optimized. An increase in frequency decreases the internal impedance of the diode-capacitor chain. However, because of the finite pulse rise time, an increase in frequency results in the pump capacitors not being discharged to the maximum which reduces the efficiency. The proposed converter has a double phase configuration as shown in Figure 3.3.4 and uses some simple logic that operates as follows. Immediately after the first pump capacitor C_{1o} has been maximally discharged, the second pump capacitor C_{1e} begins to discharge while C_{1o} recharges, and vice versa. A pump capacitor is maximally discharged if the voltage of the bottom plate has reached the supply voltage. So the chargepump circuit is made a part of an oscillator with a frequency that is determined by the size of the pump capacitors.

3.4.4 Active Stage Control

Gerber, *et al.* [4] use a method that modifies the number of active stages in the chargepump by short circuiting part of the multiplier chain. This has the advantage that the power consumption is minimized while the clock signal only charges the active capacitances. Further, a wide range of output voltages is possible. With this method the value of *N* is used as control parameter.

A method to control the value of *C* has not been found in literature. However, this can be implemented rather easily by connecting more chargepump circuits in parallel. depending on the load current and the desired output voltage. More or less parallel stage can be activated. Again, an advantage is the minimization of power consumption.

3.5 Chargepump Implementation

In this section some implementation aspects of integrated chargepump circuits are discussed. It is concluded with a description of a fully integrated chargepump circuit designed for application in the driver of a *DMOS* output stage.

3.5.1 Diodes and Switches

The simplest way to implement the diodes in a chargepump is to use isolated *pn-junctions*. However, in many *CMOS* technologies, isolated *pn-*

junctions are not available. In this case the diodes can be substituted by MOS transistors with their gate and drain contacts connected as shown in Figure 3.5.1(a). The resulting forward-bias voltage of this MOS diode is about equal to the threshold voltage V_T of the MOS transistor.

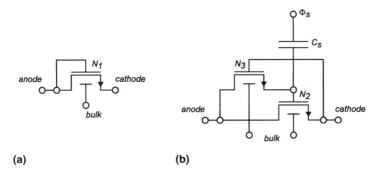

Figure 3.5.1 (a) NMOS transistor connected as a diode (b) NMOS switch configuration.

A more sophisticated solution is to use MOS switches [12,13,44] instead of diodes as shown in Figure 3.5.1(b). In this configuration, transistor N_2 is the switch. To obtain good conductivity, the gate voltage of transistor N_2 has to be at least one threshold voltage higher than the source voltage. This is achieved by using a bootstrap capacitor C_s. During the time the switch is "off", the clock Φ_s, is low. During this time the cathode voltage is at least one threshold voltage higher than the anode voltage, transistor N_3 will conduct and charge capacitor C_s up to about the anode voltage.

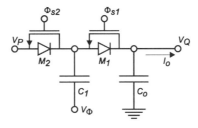

Figure 3.5.2 Chargepump using parasitic DMOS diodes

The switch is turned "on" if the anode voltage is higher than the cathode voltage. In this case, transistor N_3 is not conducting and when the clock Φ_s goes high, transistor N_2 will conduct.

A clear advantage of the use of switches instead of diodes is that switches have no forward-bias voltage drop. Consequently, the output voltage is

Chargepump Circuits

slightly higher. A disadvantage is that four clock signals are needed which require complex timing.

In case a *BCD* technology is used, it is also possible to use the parasitic diode inherent to *DMOS* transistors as shown in Figure 3.5.2. The forward-bias voltage drop can be reduced by applying a suitable gate voltage as was already mentioned in Chapter 2 [45,46]. Again additional clock signals are required.

3.5.2 Driver Circuits

In most *EEPROM* applications, simple *CMOS* inverters are used as pump drivers. The *CMOS* inverters can even be connected in a feedback loop to form a ring oscillator as shown in Figure 3.5.3 [18].

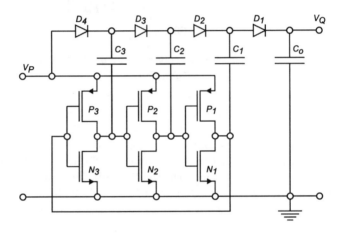

Figure 3.5.3. CMOS inverters as pump driver

In case the current driven mode is desired, current mirrors can be used. Two possibilities are shown in Figure 3.5.4. The driver shown in Figure 3.5.4(a) is suitable for fully integrated chargepumps with small current capability. In case large current capability is required, for example if the chargepump is used as a power supply, power transistors can be used in combination with external capacitors. In order to obtain high efficiency, the mirror ratio of the *DMOS* mirrors has to be chosen large (e.g. *1:1000*). For all drivers rail-to-rail operation is required to maximize the current capability and the efficiency. For the *DMOS* driver this means that the gate of the high-side mirror has to be driven with a voltage higher than V_P. This voltage can be derived from the output of the chargepump.

Chargepump Implementation

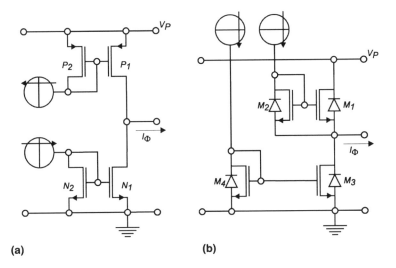

Figure 3.5.4 Current drivers for (a) low current capability (b) high current capability

3.5.3 Output Voltage Detection

In order to control the output voltage of a chargepump it is necessary to measure the output voltage. This can be a problem since this voltage is usually higher than the supply voltage. A simple solution is to use a one-sided *VI-converter* as shown in Figure 3.5.5(a).

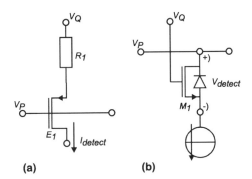

Figure 3.5.5 Output voltage detection (a) one-sided VI conversion (b) MOS transistor in triode

A disadvantage of this method is that large resistor values are needed in order to reduce the current needed by this circuit. If only a relatively small voltage difference has to be generated such as is the case with *DMOS* gate

85

Chargepump Circuits

drivers, the gate of a *DMOS* transistor can be use to measure the output voltage as shown in Figure 3.5.5(b). In this case no *DC* current is required from the chargepump output. However, it is advisable to use this circuit in combination with a Zenerlike clipping arrangement to protect the *DMOS* transistor from gate oxide damage.

3.5.4 Realization Example

A chargepump prototype has been realized in a *DMOS* process for use in a *DMOS* power output stage [47]. The pump has a double phase structure as shown in Figure 3.3.4 and Figure 3.3.8. The schematic of one of the driver circuits is shown in Figure 3.5.6. It consists of two current mirrors M_2/M_3 and E_1/E_2 operating at a supply voltage of *30V*. The current mirror E_1/E_2 is constructed with so-called extended drain *PMOS* (*EPMOS*) transistors.

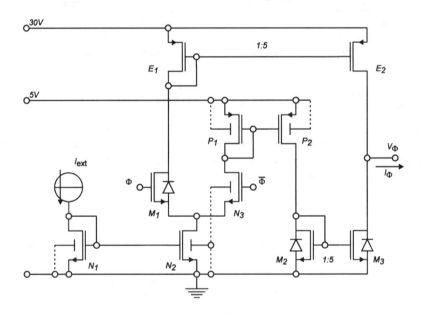

Figure 3.5.6 Chargepump driver circuit.

If the clock Φ is high then *DMOS* transistor M_1 is conducting and *EPMOS* current mirror E_1/E_2 is driven, so a current I_Φ is sourced at the output. When the clock Φ is low, transistor N_3 is conducting and *PMOS* current mirrors P_1/P_2 and *DMOS* current mirror M_2/M_3 are driven, so a current I_Φ is sunk at the output. The magnitude of the output current I_Φ is controlled by an external source I_{ext}. The chargepump can operate in both single phase and double phase mode by driving the two sides in phase or in antiphase respectively.

Both voltage driven and current driven modes can be chosen by changing the control current I_{ext}. The driver has been designed in such a way that it can deliver a current much larger than the load current in order to approach voltage driven mode. If only the current driven mode is required the driver circuit can be made much smaller. Consequently, all four operation modes described earlier can be used.

The complete chargepump circuit consists of a relaxation oscillator running at *1.67MHz*, some control logic, two driver circuits, two pump capacitors of *16.7pF* one storage capacitor of *300pF*, and four low-substrate-current diodes. The circuit requires two supply voltages V_{low} and V_{high} of *5V* and *30V* respectively. The bottom plate of the storage capacitor is connected to the power supply, so the voltage across this capacitor will be in the range of *10V*. Therefore the storage capacitor is made with thin gate oxide. The pump capacitors are charged with the full supply voltage and are therefore realized with a thick oxide.

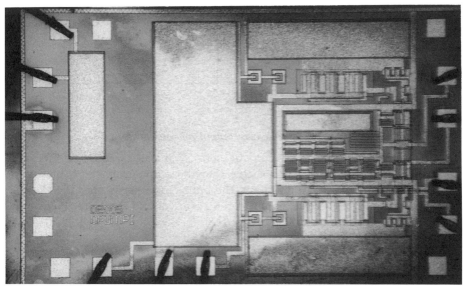

Figure 3.5.7 Chip photograph

The total area of the chargepump is about *1.2mm²*. A chip photograph is shown in Figure 3.5.7.

The measured output voltage waveforms in the four operation modes are shown in Figure 3.5.8. The load current I_o was *1.0mA* in all cases. As can be seen, the reduction in voltage ripple between the (original) voltage driven, single phase mode and the current driven, double phase mode is considerable. It is expected that a further reduction in voltage ripple is possible by improving the driver design. The fact that a rather large voltage ripple is still present is caused by two effects.

Chargepump Circuits

First, due to saturation of the current mirrors in the driver circuits, the drive currents I_ϕ will decrease if the output voltage of the driver approaches the supply rails. To compensate for this decrease, the current in the nonsaturated region has to be increased.

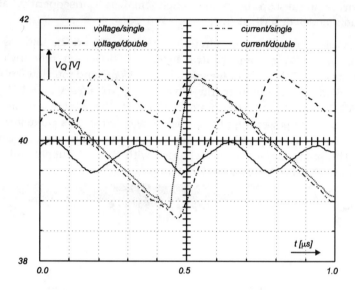

Figure 3.5.8 Measured output voltage waveforms.

Second, in this implementation the source current is equal to the sink current. Consequently it is possible that the output voltage of the driver does not reach the ground rail and the pump capacitor is not fully loaded. This reduces the effective amplitude of the driver and therefore the output voltage.

3.6 Conclusion

Voltages higher than the supply voltage can be generated with a voltage multiplier or chargepump. A chargepump does not require inductors or transformers and can therefore be integrated on-chip. An integrated chargepump is an attractive alternative for an external bootstrap capacitor.

In case a chargepump is loaded with a relatively large current, a ripple voltage appears at the output. The amplitude of this ripple voltage can be significantly reduced by applying a current driven double phase mode. A second advantage of this operation mode is that a constant current is drawn from the supply.

Several possibilities have been presented to control the output voltage of a chargepump circuit. Control is required for two reasons. First, the output voltage generated by a free running chargepump may cause damage to the circuit at the output. Second, it is often desired to have a constant output voltage under varying load conditions.

Chargepumps can also be used as a power supply. However, if large currents are required then external capacitors have to be used. Chargepump circuits cause less radiation than switching regulators with inductors, especially when the previously mentioned current driven double phase mode is used. In switching regulators a high dV/dt is necessary for high efficiency

A fully integrated chargepump has been presented that shows the feasibility of using a chargepump in a *DMOS* power stage.

3.7 References

[1] Szepesi, T., "Design and Circuit Techniques of Integrated Switching Voltage Regulator", *Proceedings of the Workshop Advances in Analog Design*, Eindhoven, The Netherlands, 29-31 Mar. 1994.

[2] Hemink, G.J., "*VIPMOS* - A Buried Injector Structure for Nonvolatile Memory Applications", Ph.D. Thesis University of Twente, 1992, ISBN 90-9005180-5.

[3] Gerber, B., J. Fellrath, "Low-Voltage Single-Supply *CMOS* Electrically Erasable Read-Only Memory", *IEEE Transactions on Electron Devices*, Vol.27, No.7, pp.1211-1216, 1980.

[4] Gerber, B., J. Martin, J. Fellrath, "A 1.5V Single-Supply One Transistor *CMOS EEPROM*", *IEEE Journal of Solid-State Circuits*, Vol.16, No.3, pp.195-200, 1981.

[5] Oto, D.H., V.K. Dham, K.H. Gudger, M.J. Reitsma, G.S. Gongwer, Y.W. Hu, J.F. Olund, H.S. Jones, S.T.K. Nieh, "High-Voltage Regulation and Process Considerations for High-Density 5V-only E^2PROMs", *IEEE Journal of Solid-State Circuits*, Vol.18, No.5, pp.532-538, 1983.

[6] Lucero, E.M., N. Challa, J. Fields, "A 16 kBit Smart 5V-only *EEPROM* with Redundancy", *IEEE Journal of Solid-State Circuits*, Vol.18, No.5, pp.539-544, 1983.

[7] McCreary, J.L., B. Eitan, D. Amrany, "Techniques for a 5V-only 64K *EPROM* based upon Substrate Hot-Electron Injection", *IEEE Journal of Solid-State Circuits*, Vol.19, No.1, pp.135-143, 1984.

[8] Yatsuda, Y., S. Nabetani, K. Uchida, S.-I. Minami, M. Terasawa, T. Hagiwara, H. Katto, T. Yasui, "*HI-MNOS* II Technology for 64-kbit Byte-Erasable 5-V-Only *EEPROM*", *IEEE Transactrions on Electron Devices*, Vol.32, No.2, pp.224-231, 1985.

[9] Yatsuda, Y., S. Nabetani, K. Uchida, S.-I. Minami, M. Terasawa, T. Hagiwara, H. Katto, T. Yasui, "*HI-MNOS* II Technology for 64-kbit Byte-Erasable 5-V-Only *EEPROM*", *IEEE Journal of Solid-State Circuits*, Vol.20, No.1, pp.144-151, 1985.

[10] D'Arrigo, S., G. Imondi, G. Santin, M. Gill, R. Cleavelin, S. Spagliccia, E. Tomassetti, S. Lin, A. Nguyen, P.Shah, G. Savarese, D. McElroy, "A 5V-Only 256k Bit *CMOS* Flash *EEPROM*", *IEEE International Solid-State Circuits Conference*, pp.132-133, February, 1989.

[11] Wijburg, R.C., G.J. Hemink, J. Middelhoek, H. Wallinga, T.J. Mouthaan, "*VIPMOS* - A Novel Buried", *IEEE Transactions on Electron Devices*, Vol.38, No.1, pp.111-120, 1991.

[12] Umewaza, A., S. Atsumi, M. Kuriyama, H. Banba, K.I. Imamiya, K. Naruke, S. Yamada, E. Obi, M. Oshikiri, T. Suzuki, S. Tanaka, "A 5-V-only Operation 0.6mm Flash *EEPROM* with Row Decoder Scheme in Triple-Well Structure", *IEEE Journal of Solid-State Circuits*, Vol.27, pp.1540-1546, 1992.

[13] Steenwijk, G. van, "Analog Applications of the *VIPMOS EEPROM*", Ph.D. Thesis University of Twente, 1994, ISBN 90-9007200-4.

Chargepump Circuits

[14] Steinhagen, W., U. Kaiser, "A Low-Power Read/Write Transponder IC for High Performance Identification Systems", *Proceedings of the 20th European Solid-State Circuits Conference*, Ulm, Germany, pp.256-259, 20-22 Sep. 1994.

[15] Storti, S., F. Consiglieri, M. Paparo, "A 30-A 30-V DMOS Motor Controller and Driver", *IEEE Journal of Solid-State Circuits*, Vol.23, No.6, pp.1394-1401, Dec. 1988.

[16] Dunn, W.C., "Driving and Protection of High-Side NMOS Power Switches", *IEEE Transactions Industrial Applications (USA)*, Vol.28, No.1, Pt.1, pp.26-30, Jan. 1992.

[17] Gariboldi, R., F. Pulvirenti, "A Monolithic Quad Line Driver for Industrial Applications", ", *ESSCIRC'93*, pp.282-285, Sep. 1993.

[18] Zitta, H., "Smart Power Circuits for Power Switches Including Diagnostic Functions", *Proceedings of the Workshop Advances in Analog Design*, Eindhoven, The Netherlands, 29-31 Mar. 1994.

[19] Vittoz, E.A., "Low-Power Low-Voltage Limitations and Prospects in Analog Design", *Proceedings of the Workshop Advances in Analog Design*, Eindhoven, The Netherlands, 29-31 Mar. 1994.

[20] Cho, T.B., D.W. Cline, C.S.G. Conroy, P.R. Gray, "Design Considerations for High-Speed Low-Power Low-Voltage CMOS Analog-to-Digital Converters", *Proceedings of the Workshop Advances in Analog Design*, Eindhoven, The Netherlands, 29-31 Mar. 1994.

[21] Dijkstra, E., O. Nys, E. Blumenkrantz, "Low Power Oversampled A/D Converters", *Proceedings of the Workshop Advances in Analog Design*, Eindhoven, The Netherlands, 29-31 Mar. 1994.

[22] Schnatz, F.V., W. Brockherde, H. Dudaicevs, B.J. Hosticka, "1.2 Volt CMOS Readout Electronics for Capacitive Sensors", *Proceedings of the 20th European Solid-State Circuits Conference*, Ulm, Germany, pp.280-283, 20-22 Sep. 1994.

[23] Nicollini, G., A. Nagari, P. Confalonieri, C. Crippa, "A -80dB THD, 4Vpp Switched Capacitor Filter for 1.5V Battery-Operated Systems", *Proceedings of the 20th European Solid-State Circuits Conference*, Ulm, Germany, pp.104-107, 20-22 Sep. 1994.

[24] Groenewold, G. "Optimal Dynamic Range Integrators", ", *IEEE Transactions on Circuits and Systems - I*, Vol.39, No.8, pp.614-627, 1992.

[25] Lin, P., L. Chua, "Topological Generation and Analysis of Voltage Multiplier Circuits", *IEEE Transactions on Circuits and Systems - I*, Vol.24, pp.517-530, Oct., 1977.

[26] Tam, K.-S., Bloodworth, E., "Automated Topological Generation and Analysis of Voltage Multiplier Circuits", *IEEE Transactions on Circuits and Systems - I*, Vol.37, No.3, Mar, 1990.

[27] Richley, E.A., "Marx Generator for High-Voltage Experiments", *Electronics & Wireless World*, Vol.93, No.1615, pp.519-523,.May 1987.

[28] Donaldson, P.E.K., "The Mosmarx Voltage Multiplier", *Electronics & Wireless World*, Vol.94, No.1630, pp.748-750, Aug. 1988.

[29] Shibata, K., M. Emura, S. Yoneda, "Energy Transmission of Switched-Capacitor Circuit and Application to DC-DC Converter", *Electronics and Communitcations in Japan, Part 2: Electronics*, Vol.74, No.4, pp.91-101, Apr. 1991.

[30] Buechel, M., "High Voltage Multipliers in TV Receivers", *IEEE Transactions on Broadcast & Television Receivers*, Vol.16, pp.32-36, 1970.

[31] Brugler, J., "Theoretical Performance of Voltage Multiplier Circuits", *IEEE Journal of Solid-State Circuits*, Vol.6, pp.132-135, June 1971.

[32] Borneman, E.H., "Boost High-Voltage DC Outputs", *IEEE Electronic Design*, pp.72-74, Mar., 1978.

[33] Malanesi, L., R. Piovan, "Theoretical Performance of the Capacitor-Diode Voltage Multiplier fed by a Current Source", *PESC'90 Record, 21st Annual IEEE Power Electronics Specialists Conference - Part II*, San Antonio, TX, USA, pp.469-476, 11-14 June, 1990.

[34] Dickson, J.F. "On-Chip High-Voltage Generation in MNOS Integrated Circuits Using an Improved Voltage Multiplier Technique", *IEEE Journal of Solid-State Circuits*, Vol.11, No.3, pp.374-378, June. 1976.

[35] Remshardt, R., H. Schettler, H. Schumacher, R. Zuehlke, "Voltage Multiplier", *IBM Technical Disclosure Bulletin*, Vol.22, No.3, pp.1052-1053, Aug. 1979.

[36] Arzubi, L., E. Baier, P. Knott, W. Loehlein, "High Performance, Low Power Voltage Doubler", *IBM Technical Disclosure Bulletin*, Vol.23, No.10, pp.4522-4524, Mar, 1981.

[37]   Durgavich, T., "Capacitive Voltage Doubler Forms ±12 to ±15V Converter", *Electronics*, Vol.55, No.2, pp.125, Jan, 1982.
[38]   Cheong, S.V., S.H. Chung, A. Ioinovici, "Duty-Cycle Control Boosts *DC-DC* Converters", *IEEE Circuits and Devices Magazine*, Vol.9, No.2, pp..36-37, 1993.
[39]   Mak, O.C., Y.C. Chung, A. Ioinovici, "Step-up *DC* Power Supply Based on a Switched-Capacitor Circuit", *IEEE Transactions on Industrial Electronics*, Vol.42, No.1, pp.90-97, Feb. 1995.
[40]   Goryunov, A.Y., "Transformerless *DC*-to-*DC* Converters", *Telecommunications and Radio Engineering*, Vol.39, pp.134-136, 1985.
[41]   Witters, J.S., G. Groeseneken, H.E. Maes, "Analysis and Modeling of On-Chip High-Voltage Generator Circuits for Use in *EEPROM* Circuits", *IEEE Journal of Solid-State Circuits*, Vol.24, No.5, pp.1372-1380, Oct. 1989.
[42]   Cataldo, G. Di, G. Palumbo, "Double and Triple Charge Pump for Power *IC*: Ideal Dynamical Models to an Optimised Design", *IEE Proceedings-G*, Vol.140, No.1, pp.33-38., Feb 1993.
[43]   Cataldo, G. Di, G. Palumbo, "Double and Triple Charge Pump for Power *IC*: Dynamic Models Which Take Parasitic Effects into Account", *IEEE Transactions on Circuits and Systems - I*, Vol.40, No.2, pp.92-101, Feb. 1993.
[44]   Steenwijk, G. van, K. Hoen, H. Wallinga, "Analysis and Design of a Charge Pump Circuit for High Output Current Applications", *ESSCIRC'93*, pp.118-121, Sep. 1993.
[45]   Chi, M.-H., C.Hu, "The Operation of Power *MOSFET* in Reverse Mode", *IEEE Transactions on Electron Devices*, Vol.30, No.12, pp.1825-1828, 1983.
[46]   Love, R.P., P.V. Gray, M.S. Adler, "A Large Area Power *MOSFET* Designed for Low Conduction Losses", *IEDM Technical Digest*, pp.418-421, 1981.
[47]   Berkhout, M., G. van Steenwijk, A.J.M. van Tuijl "A Low-Ripple Chargepump Circuit for High Voltage Applications", *ESSCIRC'95*, pp.290-293, Sep. 1995.

3.8 Appendix: Down Conversion

In some cases a *DC* voltage lower than the supply voltage is needed. Besides step-up conversion it is also possible to realize step-down converters using only capacitors and switches. In this appendix the operating principle of such step-down converters is presented together with some implementation suggestions.

3.8.1 Voltage Division

In voltage multipliers, series-parallel switching of capacitors is used to generate a multiple of the supply voltage. The same method can also be used to generate fractions of the supply voltage [48 51]. The class of circuits based on this technique is called *voltage dividers*. The operating principle of a voltage divider is illustrated in Figure 3.8.1. Capacitors C_1 and C_2 are interchanged with clock frequency f. Consequently, C_1 is connected between V_P and V_Q for half of the time and between V_Q and ground the other half of the time. For C_2 the opposite holds. Both capacitors provide part of the output current I_o. If the capacitors are equal both provide half of the output current. In first approximation it will be assumed that the switching is performed infinitely fast.

Chargepump Circuits

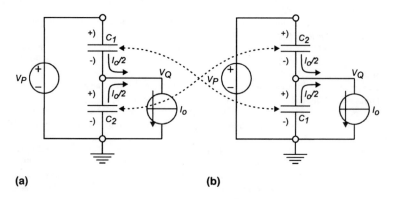

Figure 3.8.1 Voltage division

Consider the case that the output voltage is a little higher than half the supply voltage.

$$V_Q(t=0) = \frac{1}{2} \cdot V_P + \Delta V$$

(3.8.1)

If a constant load current I_o is assumed, the voltage at the output node will drop according to:

$$V_Q(t) = V_Q(0) - \frac{1}{C_1+C_2} \cdot \int_0^t I_o(\tau) d\tau = \frac{1}{2} \cdot V_P + \Delta V - \frac{I_o}{C_1+C_2} \cdot t$$

(3.8.2)

If the output voltage has dropped to $½V_P-\Delta V$, the capacitors are interchanged. Consequently the output voltage is again $½V_P+\Delta V$ returning the original situation:

$$V_Q(t) = \frac{1}{2} \cdot V_P - \Delta V \rightarrow \frac{1}{2} \cdot V_P - \Delta V = \frac{1}{2} \cdot V_P + \Delta V - \frac{I_o}{C_1+C_2} \cdot t \rightarrow$$
$$t = \frac{1}{2 \cdot f} = \frac{C_1+C_2}{I_o} \cdot 2 \cdot \Delta V$$

(3.8.3)

This results in an output waveform as shown in Figure 3.8.2. The relation between the voltage ripple V^ω_Q, frequency f and load current I_o is given by:

$$V^\omega_Q = \frac{1}{2} \cdot \frac{I_o}{f \cdot (C_1+C_2)}$$

(3.8.4)

Appendix: Down Conversion

while the steady-state output voltage V^Ω_Q of the voltage divider is given by:

$$V^\Omega_Q = \frac{1}{2} \cdot \left(V_P - V^\omega_Q\right)$$

(3.8.5)

Since the sum of the voltage over the two capacitors does not change during the switching, there will (theoretically) be no charge redistribution between the capacitors and therefore no switching losses. Therefore, the waveform shown in Figure 3.8.2 can be generated with *100%* efficiency.

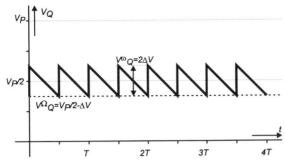

Figure 3.8.2 Output voltage waveform of the converter

However, if a constant output voltage is required, the voltage ripple has to be removed which leads to an efficiency η:

$$\eta = 2 \cdot \frac{V^\Omega_Q}{V_P}$$

(3.8.6)

This simple principle can be made even simpler by removing one of the capacitors. This results in the circuit shown in Figure 3.8.3. The remaining capacitor is connected between V_P and V_Q for half of the time and between V_Q and ground the other half. At any time the full load current is provided by the capacitor. So if the size of the conversion capacitor C_1 is equal to the sum of C_1 and C_2 in the previous circuit, the circuits are completely equivalent.

An advantage of the single-capacitor converter is that the switching involved is less complex. Further, in the inevitable case that the capacitor is to large to realize on-chip, the single-capacitor converter requires only two pins instead of the four needed in the double-capacitor converter. A possible disadvantage of the single-capacitor converter compared to the double-capacitor converter is that the supply current is not constant. Note that the difference between the single-capacitor and double-capacitor

Chargepump Circuits

version is equivalent to the single phase and double phase version used in the chargepumps.

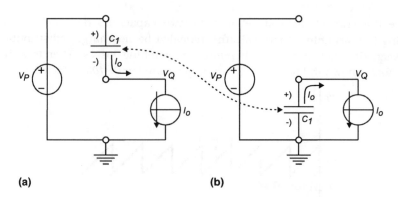

Figure 3.8.3 Single-capacitor DC step-down converter

So far it has been assumed that the switching is done infinitely fast. In a practical realization this is of course not possible which results in peaks in the output voltage. The addition of a storage capacitor C_o in parallel to the load solves this problem. This can be done without the need for extra pins since both ground and output pins are already present. If this capacitor is large enough, it can provide the load current during the switch gap. This situation is shown in Figure 3.8.4.

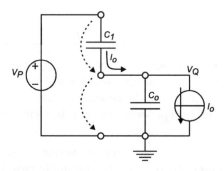

Figure 3.8.4 Converter with capacitive load

The series-parallel connection of the conversion capacitor C_1 and storage capacitor C_o will be accompanied with some energy loss due to charge redistribution. If it is assumed that the output voltage at t_0 is $V_Q(t_0)$ then after switching the output voltage $V_Q(t_1)$ is given by:

Appendix: Down Conversion

$$V_Q(t_1) = \frac{C_1}{C_o + C_1} \cdot V_P + \frac{C_o - C_1}{C_o + C_1} \cdot V_Q(t_0)$$

(3.8.7)

This equation is valid in case of switching from series to parallel but also in case of switching from parallel to series. These events are completely equivalent as far as the operation of the converter is concerned.

The effect of adding a storage capacitor on the output voltage waveform is illustrated in Figure 3.8.5. The voltage ripple V^ω_Q in this case is given by:

$$V^\omega_Q = \frac{1}{2} \cdot \frac{I_o}{f \cdot (C_1 + C_o)}$$

(3.8.8)

So the voltage ripple can be substantially reduced by adding a storage capacitor. The steady-state output voltage can be found by combining equations (3.8.7) and (3.8.8):

$$V^\omega_Q = V_Q(t_1) - V_Q(t_0) \rightarrow$$
$$V^\Omega_Q = \frac{1}{2} \cdot V_P - \frac{1}{4} \cdot \frac{I_o}{f \cdot C_1}$$

(3.8.9)

which is independent of the storage capacitor C_o. Consequently, the conversion efficiency is also unaffected by the storage capacitor.

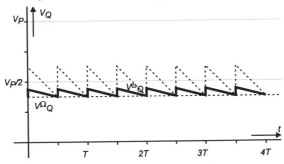

Figure 3.8.5 Effect of storage capacitor

In the chargepump circuits described in this chapter, further reduction of the voltage ripple was realized with a current driven mode. The same technique is possible in voltage dividers. If the current through the conversion capacitor C_1 is made equal to the load current then the current flowing towards the storage capacitor C_o is zero and so is the voltage ripple.

A possible realization of a current driven voltage divider is shown in Figure 3.8.6. The *DMOS* current mirrors M_1/M_2 and M_4/M_5 serve to control the

Chargepump Circuits

current through the conversion capacitor in order to reduce the voltage ripple. The mirrors have a large mirror ratio (e.g. *1:1000*) for high efficiency.

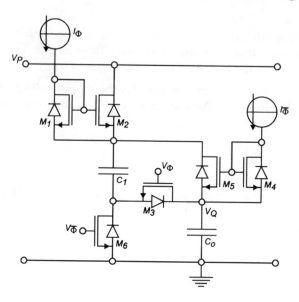

Figure 3.8.6 DMOS realization of a voltage divider

In the first half of the clock period mirror M_1/M_2 and switch M_3 are conducting. In the switch M_3 the parasitic *DMOS* diode is used to conduct current the gate drive is used to reduce the voltage drop. This situation corresponds to the situation in Figure 3.8.3(a). In the second half of the clock period mirror M_4/M_5 and switch M_6 are conducting. This situation corresponds to that shown in Figure 3.8.3(b).

3.8.2 References

[48] Oota, I., F. Ueno, T. Inoue, "Analysis of a Switched-Capacitor Transformer with a Large Voltage-Transformer-Ratio and Its Applications", *Electronics and Communications in Japan Part 2*, Vol.73, No.1, pp.85-97, 1990.

[49] Ueno, F., T. Inoue, T. Umeno, I. Oota, "Analysis and Application of Switched-Capacitor Transformers by Formulation", *Electronics and Communications in Japan Part 2*, Vol.73, No.9, pp.91-103, 1990.

[50] Shibata, K., M. Emura, S. Yoneda, "Energy Transmission of Switched-Capacitor Circuit and Application to *DC-DC* Converter", *Electronics and Communications in Japan Part 2*, Vol.74, No.4, pp.91-101, 1991.

[51] Cheong, S.V., H. Chung, A. Ioinovici, "Inductorless *DC*-to-*DC* Converter with High Power Density", *IEEE Transactions on Industrial Electronics*, Vol.41, No.2, pp.208-215, 1994.

4

Chargepump Modeling

In this chapter a detailed model that describes the operation of chargepump circuits is presented. First, a brief overview of previously published chargepump models is given. Next, the development of a new model is described that is based upon analysis of the charge balance between adjacent capacitors in the chargepump. With this model both transient and steady-state behavior of chargepumps can be described accurately. It is demonstrated that the influence of a number of parasitics can easily be included in the model.

4.1 Introduction

Most integrated chargepump circuits in use are based on the circuit proposed by Dickson in 1976 [1]. The basic electrical circuit of this chargepump is shown in Figure 4.1.1.

The circuit consists of a chain of capacitors C_n which are interconnected by diodes D_n and coupled in parallel with two non-interleaving clocksignals $\Phi(t)$ and $\overline{\Phi}(t)$ with frequency f and amplitude V_Φ. Charge is transferred from one capacitor to the next if the clock driving the first capacitor is high and the clock driving the next capacitor is low.

Chargepump Modeling

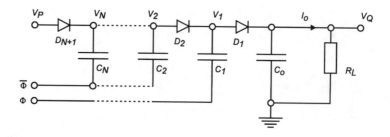

Figure 4.1.1 Basic Dickson chargepump circuit

In this section a number of previously published chargepump models is presented. In these models only part of the chargepump operation is described in detail. In the next section an extended chargepump model is developed that describes both the steady-state and dynamic behavior of a chargepump with an arbitrary number of stages.

4.1.1 Chargepump Models

A steady-state analysis of the chargepump was performed by Dickson [1]. An expression for the steady-state peak output voltage V_Q is:

$$V_Q = V_P - N \cdot \left[\left(\frac{C}{C+C_s} \right) \cdot V_\Phi - V_D \right] - V_D - \frac{N \cdot I_o}{(C+C_s) \cdot f}$$

(4.1.1)

where, C is the capacitance of the pump capacitors, C_s is the parasitic capacitance to substrate present at each node in the diode chain, V_D is the forward-bias voltage of the diodes, V_P is the supply voltage and N is the voltage multiplication factor. Periodic charging and discharging of the output capacitor C_o causes a voltage ripple at the output node. The contribution to the ripple is caused by the current through the load resistor R_L. This contribution V_R can be calculated by:

$$V_R = \frac{I_o}{f \cdot C_o} = \frac{V_Q}{f \cdot R_L \cdot C_o}$$

(4.1.2)

A second contribution is crosstalk from the clock V_Φ due to the parasitic parallel capacitance C_D of the diodes. This contribution V_C can be calculated with:

Introduction

$$V_C = \frac{C_D}{C_o + C_D} \cdot \frac{C}{C + C_S} \cdot V_\Phi$$

(4.1.3)

The diodes have a non-zero internal series resistance R_D which causes the output resistance of the chargepump to increase non-linearly with the load current. Further, it is pointed out that the threshold voltages of the diode-connected MOS transistors are not equal due to the body-effect. A mean threshold value is assumed to account for this.

Witters et al. [2] extended Dickson's steady-state model by considering the influence of the body-effect in more detail. The body-effect is approximated by a parameter α in the following relationship:

$$V_s = tan(\alpha) \cdot (V_d - V_{t0})$$

(4.1.4)

where, V_s and V_d are the source and drain voltages of a conducting MOS diode and V_{t0} is the zero-bias threshold voltage. An expression for the steady-state peak output voltage including the body-effect is found to be:

$$V_Q = tan^{n+1}(\alpha) \cdot V_P - tan(\alpha) \cdot \frac{1 - tan^{n+1}(\alpha)}{1 - tan(\alpha)} \cdot V_{t0} + \left(\frac{C}{C + C_S} \cdot V_\Phi - \frac{I_{out}}{f \cdot (C + C_S)} \right) \cdot tan(\alpha) \cdot \frac{1 - tan^n(\alpha)}{1 - tan(\alpha)}$$

(4.1.5)

It is pointed out that the accuracy of this model is limited due to the linear approximation of the body-effect which is only valid under certain conditions. Further, it is mentioned that leakage currents of any nature can cause deviations.

Van Steenwijk et al. [3,4] analyzed the equivalent output resistance R_{out} of the chargepump in case MOS switches are used instead of diodes In this analysis, the internal series resistance R_{on} of the switches is taken into account. The author presents a chargepump circuit that uses PMOS transistors in separate n-wells to implement switches. The body-effect is avoided by keeping the source-well voltage constant. The analysis results in the following expression for the equivalent output resistance:

$$R_{out} = \frac{N}{f \cdot C} coth\left(\frac{2}{R_{on} \cdot f \cdot C} \right)$$

(4.1.6)

A transient analysis of a double and a triple chargepump was performed by Di Cataldo et al. [5]. With dynamic models it is possible to calculate the

Chargepump Modeling

starup behavior of the double and triple chargepump using simple expressions:

$$V_{Q,double}(nT) = V_P - 2 \cdot V_D + \left[1 - \left(\frac{K}{1+K}\right)^n\right] \cdot V_\Phi$$

(4.1.7)

$$V_{Q,triple}(nT) = V_P - 3 \cdot V_D + 2 \cdot \left[1 - \left(\frac{1+4 \cdot K}{2+4 \cdot K}\right)^n\right] \cdot V_\Phi$$

(4.1.8)

where n is the number of clock cycles, K is equal to C_o/C in the double chargepump case and equal to $C_o/2C$ in the triple chargepump case.

In a later publication [6], the influence of parasitic capacitance to substrate and load current were taken into account.

4.2 Extended Chargepump Model

In this section an extended chargepump model is presented. It is shown that the model can be used to calculate both the steady-state and the transient behavior of chargepumps with an arbitrary number of stages. The basic model assumes ideal circuit components with no parasitics. However, the influence of parasitic effects can be easily included in both steady-state and transient equations as is demonstrated in Section 4.3. The model can also be used to calculate power efficiency during steady-state and transients as is demonstrated in Section 4.4.

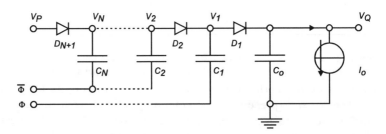

Figure 4.2.2 N-stage *chargepump model*

Furthermore, it is shown that the steady-state and transient models described previously all derive from this model.

4.2.1 Some Notation Conventions

An *N-stage* chargepump consists of N so called pump capacitors and one storage capacitor connected by diodes as shown in Figure 4.2.2.

The pump capacitors are driven by two non-overlapping clocksignals Φ and $\overline{\Phi}$ both having amplitude V_Φ. The clock Φ drives the nodes with odd node numbers and the inverted clock $\overline{\Phi}$ drives the nodes with even node numbers. All charge transfer is assumed to be instantaneous. Consequently, the state of the chargepump can be calculated by calculating the charge balance before and after each clock transition. The chargepump is loaded with a current $I_o(k)$ which is assumed to be constant during each clock cycle k.

Clock Signals

The clock signals used are shown in Figure 4.2.3. The first half of each clock cycle k will be called the *odd phase* Φ_o while the second will be called the *even phase* Φ_e. The amplitude of the clock signals $V_\Phi(k)$ can change between clock cycles but is assumed to be constant during each clock cycle.

The pump capacitor preceding the storage capacitor C_o is assumed to be connected to clock Φ. Consequently, all charge transfer towards the storage capacitor will take place at the beginning of each clock cycle k. During the remainder of the clock cycle, the storage capacitor is loaded with a constant current $I_o(k)$ which causes the output voltage to drop. So, the output voltage has a minimum level at the end of each clock cycle.

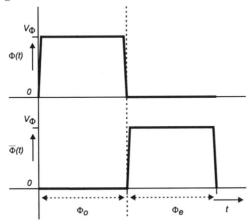

Figure 4.2.3 Clock signals

From this point on the subscripts $2m$, $2m+1$ and $2m-1$ are used to distinguish between *odd* and *even* nodes while the subscript n is used to

Chargepump Modeling

indicate any node. Further, the subscripts Φ_o and Φ_e are used to distinguish between the odd and even *clock phases*. So, for example the term $V_{2m}(k)_{\Phi o}$ indicates the voltage on the even node with node number *2m* at the end of the *odd phase* of the *k-th* clock cycle.

Ground State

The capacitors are interconnected by diodes D_n which are assumed to be ideal, that is $I_d=0$ if $V_d<V^\delta_n$ and $V_d=V^\delta_n$ if $I_d>0$. The voltage V^δ_n is the forward-bias voltage of diode D_n. In some applications, the diodes are realized with diode-connected *MOS* transistors. In this case the forward-bias voltage is equal to the threshold voltage of the *MOS* transistors which depends on the source voltage due to the body-effect. Therefore, it is assumed that the forward-bias voltage V^δ_n can have different values for different diodes D_n. If the clocksignals are inactive it can easily be seen that after some time the node voltages along the diode chain are given by:

$$V_n^\Gamma = V_P - \sum_{k=n+1}^{N+1} V_k^\delta$$

(4.2.9)

and are independent of the load current $I_o(k)$. This situation will be called the *ground state* Γ of the chargepump. Clearly, the voltage on any of the nodes will never drop below its ground state value.

State Variables

If all charge transfers are assumed to be instantaneous, the state of the chargepump at each half clock cycle can be described by the difference between the voltage $V_n(k)$ on that node and the ground state voltage V^Γ_n on that same node:

$$\Delta V_n(k) = V_n(k) - V_n^\Gamma$$

(4.2.10)

The number of state variables can be reduced by recognizing that only the state at the end of each clock cycle *k*, which is the minimal value, is important. This means that the chargepump behavior can be described by only considering the voltages at the end of each clock cycle, that is, the even phase Φ_e. Further it can easily be seen that in the even phase, there is a relation between the odd node and even node voltages. In the even phase, the clock Φ, driving the odd nodes, is low while the clock $\overline{\Phi}$, driving the even nodes, is high. So the diodes connecting the odd nodes *2m-1* to the preceding even nodes *2m* will be forward-biased and the voltage difference between these nodes will be the forward-bias diode voltage V^δ_{2m} of the connecting diode D_{2m}. Since the voltage difference between the respective ground states is the V^s_{2m} of the same diode D_{2m} this means that their difference voltage is the same:

$$\Delta V_{2m}(k)_{\Phi_e} = \Delta V_{2m-1}(k)_{\Phi_e} \qquad (4.2.11)$$

Consequently, the state of the chargepump can be described completely by the voltage differences on the *even* nodes at the end of the even clock phase Φ_e.

4.2.2 Transient Modeling

In this section a dynamic model for a *N-stage* chargepump is developed.

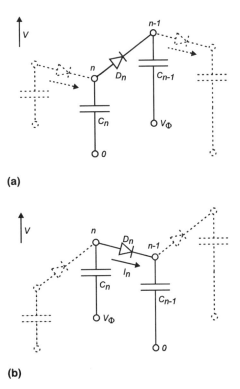

Figure 4.2.4 Charge Balance

The dynamic behavior of the chargepump can be described with the following *M-dim*ensional difference equation:

$$\Delta V(k) = A \cdot \Delta V(k-1) + B \cdot u(k) \qquad (4.2.12)$$

Chargepump Modeling

where $\Delta V(k)$ is the state vector consisting of the even node difference voltages at the end of each clock cycle, A is the system matrix, B is the input matrix and u is the input vector which depends on the load current $I_o(k)$ and clock amplitude $V_\Phi(k)$. The dimension M is equal to the number of even nodes in the chargepump, including the output node.

Charge Balance Equations

The matrices A and B and the vector u can be determined by considering the charge balance on the chargepump nodes before and after clock transitions. In the situation shown in Figure 4.2.4 the total charge stored on capacitors C_n and C_{n-1} does not change as a result of the clock transition. However, the distribution of charge does change since a part of the charge on C_n is transferred to C_{n-1} through the connecting diode D_n.

In the even phase charge is transferred from all even nodes to the succeeding odd nodes the charge balance results in:

$$q_{2m-1}(k)_{\Phi_e} + q_{2m}(k)_{\Phi_e} = q_{2m-1}(k)_{\Phi_o} + q_{2m}(k)_{\Phi_o} \qquad (4.2.13)$$

Then substituting:

$$\begin{cases} q_{2m}(k)_{\Phi_e} = C_{2m} \cdot \left(V_{2m-1}(k)_{\Phi_e} + V^\delta_{2m} - V_\Phi(k)\right) \\ q_{2m-1}(k)_{\Phi_e} = C_{2m-1} \cdot V_{2m-1}(k)_{\Phi_e} \\ q_{2m}(k)_{\Phi_o} = C_{2m} \cdot V_{2m}(k)_{\Phi_o} \\ q_{2m-1}(k)_{\Phi_o} = C_{2m-1} \cdot \left(V_{2m-1}(k)_{\Phi_o} - V_\Phi(k)\right) \end{cases} \qquad (4.2.14)$$

into equation (4.2.13) leads to:

$$\begin{aligned} V_{2m-1}(k)_{\Phi_e} &= \frac{C_{2m-1}}{C_{2m-1}+C_{2m}} \cdot \left(V_{2m-1}(k)_{\Phi_o} - V_\Phi(k)\right) \\ &+ \frac{C_{2m}}{C_{2m-1}+C_{2m}} \cdot \left(V_{2m}(k)_{\Phi_o} - V^\delta_{2m} + V_\Phi(k)\right) \\ &= \theta_{2m-1_{\Phi_e}} \cdot \left(V_{2m}(k)_{\Phi_o} - V_\Phi(k)\right) \\ &+ \theta_{2m_{\Phi_e}} \cdot \left(V_{2m}(k)_{\Phi_o} - V^\delta_{2m} + V_\Phi(k)\right) \end{aligned} \qquad (4.2.15)$$

for the odd node voltages and,

Extended Chargepump Model

$$V_{2m}(k)_{\Phi_e} = V_{2m-1}(k)_{\Phi_e} + V_{2m}^{\delta}$$

(4.2.16)

for the even node voltages. For the odd phase a similar calculation can be done. For the output node of the chargepump, the charge balance equation is slightly different to include the charge difference caused by the load current I_o.

State Transition Equations

Next, the state variables $\Delta V_n(t)$ are introduced. Equation (4.2.10) is rewritten as:

$$V_n(k)_{\Phi_o} = V_n^{\Gamma} + \Delta V_n(k)_{\Phi_o}$$
$$V_n(k)_{\Phi_e} = V_n^{\Gamma} + \Delta V_n(k)_{\Phi_e}$$

(4.2.17)

and, from equation (4.2.9) can be derived that:

$$V_n^{\Gamma} = V_{n+1}^{\Gamma} - V_{n+1}^{\delta}$$

(4.2.18)

Substitution of equations (4.2.17) and (4.2.18) in the voltage equations results in the *state transition equations*:

$$\Delta V_{2m}(k)_{\Phi_o} = \theta_{2m\Phi_o} \cdot \left(\Delta V_{2m}(k-1)_{\Phi_e} - V_{\Phi}(k-1)\right)$$
$$+ \theta_{2m+1\Phi_o} \cdot \left(\Delta V_{2m+2}(k-1)_{\Phi_e} + V_{\Phi}(k)\right)$$

(4.2.19)

for the odd phase and,

$$\Delta V_{2m}(k)_{\Phi_e} = \theta_{2m-1\Phi_e} \cdot \left(\Delta V_{2m-2}(k)_{\Phi_o} - V_{\Phi}(k)\right)$$
$$+ \theta_{2m\Phi_e} \cdot \left(\Delta V_{2m}(k)_{\Phi_o} + V_{\Phi}(k)\right)$$

(4.2.20)

for the even phase. In matrix form this leads to:

$$\Delta V(k)_{\Phi_o} = A_{\Phi_o} \cdot \Delta V(k-1)_{\Phi_e} + B_{\Phi_o} \cdot u(k)$$
$$\Delta V(k)_{\Phi_e} = A_{\Phi_e} \cdot \Delta V(k)_{\Phi_o} + B_{\Phi_e} \cdot u(k)$$

(4.2.21)

where $\Delta V(k)_{\Phi o}$ and $\Delta V(k)_{\Phi e}$ are vectors containing the state variables:

Chargepump Modeling

$$\Delta V(k)_{\Phi_o} = \begin{pmatrix} \Delta V_0(k)_{\Phi_o} & \Delta V_2(k)_{\Phi_o} & \Delta V_4(k)_{\Phi_o} & \cdots \end{pmatrix}^T$$

$$\Delta V(k)_{\Phi_e} = \begin{pmatrix} \Delta V_0(k)_{\Phi_e} & \Delta V_2(k)_{\Phi_e} & \Delta V_4(k)_{\Phi_e} & \cdots \end{pmatrix}^T$$

(4.2.22)

and $u(k)$ is the input vector:

$$u(k) = \begin{pmatrix} I_o(k)/fC_1 & V_\Phi(k) & V_\Phi(k-1) \end{pmatrix}^T$$

(4.2.23)

The difference equation (4.2.12) can now be found by eliminating $\Delta V(k)_{\Phi o}$ from (4.2.21). The system matrix A and input matrix B are then equal to:

$$A = A_{\Phi_e} \cdot A_{\Phi_o}$$
$$B = B_{\Phi_e} + A_{\Phi_e} \cdot B_{\Phi_o}$$

(4.2.24)

The dynamic equation (4.1.4) can be used to calculate the response of the chargepump to arbitrary changes in load current and clock amplitude. The response of the system on an input signal $u(k)$ is given by:

$$\Delta V(k) = A^k \cdot \Delta V(0) + \sum_{l=0}^{k-1} A^l \cdot B \cdot u(k-1-l)$$

(4.2.25)

If a step in the input occurs, the system will converge to a new steady-state output voltage V^Ω_Q according to:

$$V_Q(k) = V_Q^\Gamma + \Delta V_Q^\Omega(0) + \left(\Delta V_Q^\Omega(\infty) - \Delta V_Q^\Omega(0)\right) \cdot \left(1 - e_1 \cdot \lambda_1^k - \ldots - e_M \cdot \lambda_M^k\right)$$

(4.2.26)

where λ_n are the eigenvalues of the system matrix A. If the load current $I_o(k)=0$ and the clock amplitude $V_\Phi(k)=V_\Phi$, then for $N=1$ and $N=2$, equation (4.2.26) is equal to the double and triple chargepump models published by Di Cataldo et al. For clarity, in this model the parameter N is the number of pump capacitors.

It should be noted that the model implicitly assumes that no two successive diodes are conducting at the same time. If the diodes are replaced with switches this is always true. If real pn-diodes or MOS diodes are used this assumption is not always correct.

4.2.3 Steady-state Modeling

If the load current I_o and clock amplitude V_Φ are kept constant, the chargepump will converge towards a *steady-state* V^Ω. The steady-state output voltage V^Ω_Q of a chargepump can be derived from the difference equation (4.2.12):

$$\Delta V^\Omega = (I - A)^{-1} \cdot B \cdot u \qquad (4.2.27)$$

where I is the identity matrix of the same dimension as A. The steady-state output voltage is then given by:

$$V^\Omega_Q = V^\Gamma_Q + \Delta V^\Omega_Q \qquad (4.2.28)$$

Steady-state Output Voltage

A simple expression for the steady-state output voltage V^Ω_Q can be found if it is assumed that all pump capacitors have equal size $C_n=C$, the storage capacitor has size $C_o=\kappa C$ and all diodes have the same forward-bias voltage $V^\delta_n=V_\Delta$. If the clock driving capacitor C_n is low the voltage on node n is equal to the voltage of the preceding node $n+1$ minus V_Δ. If the clock rises the initial voltage rise on node n is equal to V_Φ. Then an amount of charge equal to I_o/f is transferred to the succeeding capacitor C_{n-1}. This results in a voltage drop of I_o/fC on node n. So the contribution of all stages to the output voltage equals:

$$V_P + \sum_{n=1}^{N}\left(-V_\Delta + V_\Phi - \frac{I_o}{f \cdot C}\right) = V_P - N \cdot V_\Delta + N \cdot V_\Phi - N \cdot \frac{I_o}{f \cdot C} \qquad (4.2.29)$$

At the start of the even phase, the output node voltage is equal to the voltage at node 1 minus V_Δ. Then the voltage drops an additional $(I_o/2)/f\kappa C$ due to the load current I_o. Consequently, the steady-state output voltage V^Ω_Q is given by:

$$V^\Omega_Q = V_P - (N+1) \cdot V_\Delta + N \cdot V_\Phi - \left(N + \frac{1}{2 \cdot \kappa}\right) \cdot \frac{I_o}{f \cdot C}$$

$$= V^\Gamma_Q + N \cdot V_\Phi - \left(N + \frac{1}{2 \cdot \kappa}\right) \cdot \iota_o \qquad (4.2.30)$$

where ι_o is defined as the effective load current I_o/fC. This expression is similar to the steady-state expression derived by Dickson. However, equation (4.2.30) gives the minimal output voltage instead of the peak

Chargepump Modeling

output voltage. This explains the appearance of the factor κ in the steady-state output voltage.

Output Voltage Ripple

The output node voltages that result from the general chargepump model are only valid at the end of each clock cycle. However, the output voltage is not constant during the clock cycle but changes as a result of the charge transfer and load current. During the clock edge between the even and odd clock phase, charge is transferred from C_1 to the storage capacitor C_o causing the output voltage to rise instantaneously. In the development of the chargepump model it was assumed that the load current is constant during each clock cycle. In this case a total charge of $Q(k)=I_o(k)/f$ is drained from the output capacitor C_o during each clock cycle k. Half of this charge $Q(k)$ is drained during the odd clock phase while the other half is drained during the even clock phase. In the odd clock phase, the clock driving pump capacitor C_1 is high and diode D_1 connecting C_1 and C_o is forward biaseded thus connecting C_o and C_1 in parallel. Consequently the voltage drops linearly with time:

$$\frac{\partial}{\partial t}V_Q(t) = -\frac{1}{C_o + C_1} \cdot I_o$$

(4.2.31)

Next, during the even clock phase, the second half of the charge $Q(k)$ is withdrawn from the output. However, since the clock driving C_1 is low, diode D_1 is reverse-biased and C_1 is isolated from C_o.

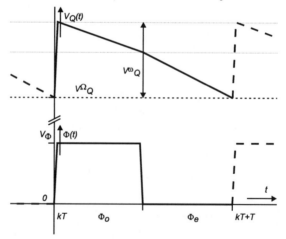

Figure 4.2.5 Output voltage waveform

In this phase the voltage drop is:

$$\frac{\partial}{\partial t} V_Q(t) = -\frac{1}{C_o} \cdot I_o$$

(4.2.32)

So, the voltage drop in the odd clock phase is smaller than in the even clock phase. In practical implementations C_o will be much larger than C_1 so this difference will be very small.

If the pump is in steady-state then the output voltage at the end of each clock cycle k is equal, i.e. the instantaneous output voltage rise that occurs during the rising edge of clock Φ is equal to the voltage drop during the odd and even phases of the subsequent clock period. The output voltage waveform is shown in Figure 4.2.5. In this case the output voltage ripple V^{ω}_Q is given by:

$$V^{\omega}_Q = \frac{I_o}{2 \cdot f} \cdot \left(\frac{1}{C_o} + \frac{1}{C_o + C_1} \right) \approx \frac{I_o}{f \cdot C_o}$$

(4.2.33)

This voltage ripple V^{ω}_Q can not be considered as a parasitic effect since it is inherent to the chargepump operation.

4.2.4 Transient Behavior

The influence of various parameters on the start-up behavior of chargepump circuits is considered here in more detail. the number of clock cycles needed to reach a new steady-state V^{Ω}_Q is only dependent on the system matrix A introduced in equation (4.1.4) which is determined by the number of stages N and capacitor ratios. The *ground state* V^r_Q, supply voltage V_P, clock amplitude V_Φ, forward-bias diode voltage V_Δ, and effective load current ι_o only influence the steady-state V^{Ω}_Q that the output of the chargepump converges to.

Usually, both the supply voltage V_P and clock amplitude V_Φ are equal to the available supply voltage and the forward-bias diode voltage V_Δ is a constant determined by the technology used. Therefore, only the values of N, κ and ι_o can be chosen freely.

The influence of the number of stages N is shown in Figure 4.2.6. The other values used in these calculations were: $f=1MHz$, $I_o=1.0mA$, $C=1nF$, $(\iota_o=1)$, $\kappa=9$, $V_P=V_\Phi=10V$ and $V_\Delta=0.5V$. For comparison all voltages have been normalized by dividing them by V_P+NV_Φ.

As can be seen in Figure 4.2.6 the settling time of a chargepump decreases rapidly when the number of stages increases.

Chargepump Modeling

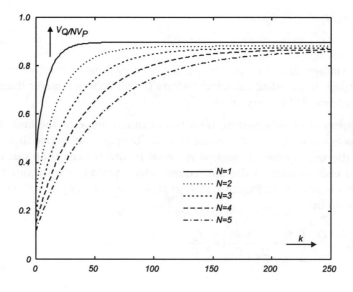

Figure 4.2.6 Normalized dynamic chargepump output for different values of N

In Figure 4.2.7 the normalized output voltage of a *3-stage* chargepump is shown for different values of the ratio κ, i.e. different values for the storage capacitor C_o.

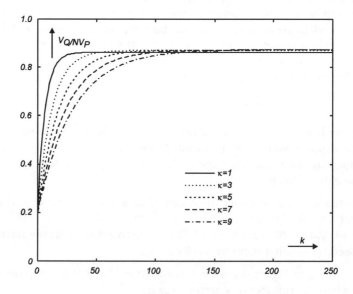

Figure 4.2.7 Normalized dynamic chargepump output for different values of κ

Modeling of Parasitic Effects

The values of the other parameters are equal to the previous case. From Figure 4.2.7 it can be seen that a larger value of κ leads to a longer settling time and also to a slightly higher steady-state output voltage which results from a smaller voltage ripple.

Finally, in Figure 4.2.8 the normalized output voltage of a *3-stage* chargepump for different values of the effective load current ι_o is shown. Again, the values of the other parameters is the same as previously. The load current does not influence the settling time of the chargepump but does result in a lower steady-state value.

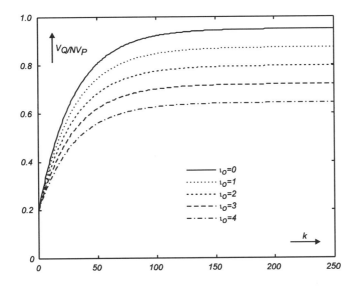

Figure 4.2.8 Normalized dynamic chargepump output for different values of ι_o

All calculations presented here have been compared with circuit simulations with idealized components and the matching was very accurate.

4.3 Modeling of Parasitic Effects

In the development of the general chargepump model described in the previous section, it is assumed that all components of the chargepump have a very simple model. Of course, in practice this is not the case and the behavior of the chargepump will be influenced by parasitic effects. In order to estimate the influence of each of these parasitics on both the transient and the steady-state behavior of the chargepump, they are analyzed separately in the following sections.

Chargepump Modeling

4.3.1 Current Leakage

The diodes used in the chargepump can be realized as isolated diodes, MOS transistors connected as diodes, or switches. In all cases there will be some junction leakage current to the substrate. To estimate the effect of current leakage on the chargepump behavior it is assumed that the leakage current is proportional to the diode current with ratio λ.

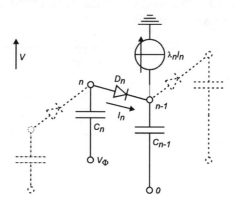

Figure 4.3.1 Current leakage

In circuit simulations, this parasitic effect can be modeled with a controlled current source at the cathode of the diodes as shown in Figure 4.3.1. Due to the current leakage, some charge is lost during every charge transfer in the chargepump.

Transient Behavior

The current leakage appears as an additional term q^λ in the charge balance. In the even phase, for example, this leads to:

$$q_{2m-1}(k)_{\Phi_e} + q_{2m}(k)_{\Phi_e} + q_{2m}^\lambda(k) = q_{2m-1}(k)_{\Phi_o} + q_{2m}(k)_{\Phi_o}$$

(4.3.1)

The amount of leaked charge is proportional to the amount of charge transferred:

$$q_{2m}^\lambda(k) = \lambda_{2m} \cdot \left(q_{2m}(k)_{\Phi_o} - q_{2m}(k)_{\Phi_e} \right)$$

(4.3.2)

Substitution of (4.3.2) into (4.3.1) leads to:

Modeling of Parasitic Effects

$$q_{2m-1}(k)_{\Phi_e} + q_{2m}(k)_{\Phi_e} \cdot (1-\lambda_{2m}) = q_{2m-1}(k)_{\Phi_o} + q_{2m}(k)_{\Phi_o} \cdot (1-\lambda_{2m})$$
(4.3.3)

From this point the model development is similar to that of the parasitic free case and leads to an adapted system matrix A^λ and input matrix B^λ.

The effect of current leakage on the start-up behavior of a *3-stage* chargepump for different values of λ is shown in Figure 4.3.2. The bottom curve shows the predicted start-up together with a circuit simulation. In the inset it can be seen that model and simulation match accurately.

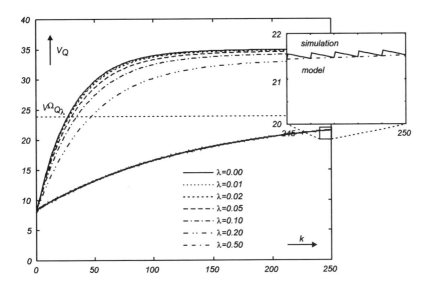

Figure 4.3.2 Influence of current leakage on start-up behavior

Steady-state Output Voltage

An expression for the steady-state output voltage $V^\Omega_{Q\lambda}$ can be obtained under the same assumptions as for the ideal steady-state expression, i.e. all pump capacitors have equal size $C_n=C$, the storage capacitor has size $C_o=\kappa C$ and all diodes have the same forward-bias voltage $V^\delta_n=V_\Delta$. Further, it is assumed that all diodes have the same current leakage parameter value $\lambda_n=\lambda$. If the clock driving capacitor C_n is low, the voltage on node n is equal to the voltage of the preceding node $n+1$ minus V_Δ. If the clock rises, the initial increase in the node voltage is equal to V_Φ. Then an amount of charge equal to $I_o/f(1-\lambda)^n$ is transferred to the succeeding capacitor C_{n-1} resulting in a voltage drop $I_o/fC(1-\lambda)^n$. Consequently, the contribution of all stages equals:

Chargepump Modeling

$$\sum_{n=1}^{N}\left(V_\Phi - V_\Delta - \frac{I_o}{f\cdot C}\cdot\frac{1}{(1-\lambda)^n}\right) = N\cdot V_\Phi - N\cdot V_\Delta - \frac{1-(1-\lambda)^N}{\lambda\cdot(1-\lambda)^N}\cdot\frac{I_o}{f\cdot C}$$

(4.3.4)

At the start of the even phase, the output node voltage is equal to the voltage at node 1 minus V_Δ. Then the voltage drops an additional $(I_o/2)/f\kappa C$ due to the load current I_o. This leads to the steady-state output voltage:

$$V^\Omega_{Q_\lambda} = V_P - (N+1)\cdot V_\Delta + N\cdot V_\Phi - \left(\frac{1-(1-\lambda)^N}{\lambda\cdot(1-\lambda)^N} + \frac{1}{2\cdot\kappa}\right)\cdot I_o$$

(4.3.5)

The output voltage ripple is not affected by current leakage.

As can be seen from Figure 4.3.2 and (4.3.5) the presence of leakage currents results in both a lower steady-state output voltage and a longer settling time. Due to the leakage, charge is lost via all forward biased diodes. In the case of substrate isolated diodes this current is linearly proportional to the diode current for a number of decades. In practice this leakage current will be a few orders of magnitude smaller than the diode current. Consequently, the influence of current leakage can be considered negligible.

4.3.2 Series Resistance

The series resistance that is always present in diodes, MOS diodes and switches, influences the charge transfer from one capacitor to the next. In circuit simulations the series resistance is modeled as a resistor in series with the diodes as shown in Figure 4.3.3.

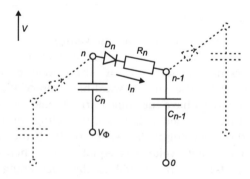

Figure 4.3.3 Series resistance

Modeling of Parasitic Effects

If the series resistance becomes high, the charge transfer cannot be considered instantaneous anymore but will exhibit an exponential relaxation in time. The time constant τ of this relaxation is equal to the product of the series resistance R_n and the capacitors at the anode and cathode of the diode D_n in parallel. If the relaxation time constant is not small compared to the clock cycle, the current through R_n will not be zero when the next clock transition occurs and consequently there will be a nonzero voltage drop over the series resistance R_n that is proportional to the amount of charge transferred.

Transient Behavior

Immediately after the clock transition the voltages on the nodes n and $n-1$ are $V_n(0)$ and $V_{n-1}(0)$ respectively. The node voltages as a function of time are found by solving the differential equations:

$$C_n \cdot \frac{\partial V_n(t)}{\partial t} = -I_n(t)$$

$$C_{n-1} \cdot \frac{\partial V_{n-1}(t)}{\partial t} = I_n(t)$$

$$V_n(t) - V_{n-1}(t) = V_n^\delta - I_n(t) \cdot R_n$$

(4.3.6)

which leads to:

$$V_n(t) = \frac{C_{n-1} \cdot (V_{n-1}(0) + V_n^\delta) + C_n \cdot V_n(0)}{C_n + C_{n-1}} + \frac{C_{n-1}}{C_n + C_{n-1}} \cdot V_n^p(t)$$

$$V_{n-1}(t) = \frac{C_{n-1} \cdot V_{n-1}(0) + C_n \cdot (V_n(0) - V_n^\delta)}{C_n + C_{n-1}} - \frac{C_n}{C_n + C_{n-1}} \cdot V_n^p(t)$$

$$V_n^p(t) = V_n(t) - V_{n-1}(t) - V_n^\delta = \left(V_n(0) - V_{n-1}(0) - V_n^\delta\right) \cdot e^{-\frac{t}{\tau_n}}$$

(4.3.7)

where the time constant τ_n is given by:

$$\tau_n = R_n \cdot \frac{C_n \cdot C_{n-1}}{C_n + C_{n-1}}$$

(4.3.8)

After half a clock cycle the voltage across the resistor is given by:

$$V_n^p\left(\frac{1}{2f}\right) = \left(V_n(0) - V_{n-1}(0) - V_n^\delta\right) \cdot e^{-\frac{t}{2 \cdot f \cdot \tau_n}}$$

(4.3.9)

Chargepump Modeling

The charge balance remains unchanged since no charge is lost during transfer. So in the even phase:

$$q_{2m}(k)_{\Phi_e} + q_{2m-1}(k)_{\Phi_e} = q_{2m}(k)_{\Phi_o} + q_{2m-1}(k)_{\Phi_o}$$

(4.3.10)

The nonzero voltage across the resistor can simply be added to the forward-bias voltage drop of the diode:

$$\begin{cases} q_{2m}(k)_{\Phi_e} = C_{2m} \cdot \left(V_{2m-1}(k)_{\Phi_e} + V_{2m}^{\delta} + V_{2m}^{\rho}(k)_{\Phi_e} - V_{\Phi}(k)\right) \\ q_{2m-1}(k)_{\Phi_e} = C_{2m-1} \cdot V_{2m-1}(k)_{\Phi_e} \\ q_{2m}(k)_{\Phi_o} = C_{2m} \cdot V_{2m}(k)_{\Phi_o} \\ q_{2m-1}(k)_{\Phi_o} = C_{2m-1} \cdot \left(V_{2m-1}(k)_{\Phi_o} - V_{\Phi}(k)\right) \end{cases}$$

(4.3.11)

Due to the relaxation behavior, this residual voltage is proportional to the initial voltage difference over the resistor. Using equation (4.3.7) it follows that:

$$V_{2m}^{\rho}(k)_{\Phi_e} = \rho_{2m} \cdot \left(V_{2m}(k)_{\Phi_o} + V_{\Phi}(k) - V_{2m-1}(k)_{\Phi_o} + V_{\Phi}(k) - V_{2m}^{\delta}\right)$$

(4.3.12)

where the parameter ρ is defined as:

$$\rho_n = e^{-\frac{1}{2 \cdot f \cdot \tau_n}}$$

(4.3.13)

From here on the model development is similar to that of the parasitic free case and leads to an adapted system matrix A^ρ and input matrix B^ρ. In this case the dimension of the difference equation that describes the dynamic behavior of the chargepump is equal to the number of nodes $N+1$. Due to the series resistance, the voltage between subsequent nodes is not always equal to one forward-bias diode voltage but depends on the load current $I_o(k)$. Consequently, the reduction in number of state variables is not allowed here.

The influence of series resistance on the start-up behavior of a *3-stage* chargepump is shown in Figure 4.3.4. The bottom curve shows the predicted start-up together with a circuit simulation. In the inset it can be seen that model and simulation match accurately.

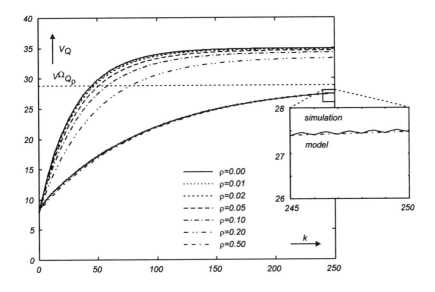

Figure 4.3.4 Influence of diode series resistance on start-up behavior.

Steady-state Output Voltage

An expression for the steady-state output voltage V^{Ω}_{Qp} can be obtained under the same assumptions as for the ideal steady-state expression, i.e. all pump capacitors have equal size $C_n=C$, the storage capacitor has size $C_o=\kappa C$ and all diodes have the same forward-bias voltage $V^{\delta}_n=V_\Delta$. Further, it is assumed that all diodes have the same series resistance parameter value $\rho_n=\rho$. If the clock driving capacitor C_n is low then the voltage at node n is equal to the voltage of the preceding node $n+1$ minus V_Δ minus one residual voltage V^{ρ}_{n+1}. If the clock rises the initial voltage rise on node n is equal to V_Φ. Then an amount of charge equal to I_o/f is transferred to the succeeding capacitor C_{n-1}. This results in a voltage drop of I_o/fC on node n. So the contribution of all stages to the output voltage equals:

$$V_P + \sum_{n=1}^{N}\left(-V_\Delta - V^{\rho}_{n+1} + V_\Phi - \frac{I_o}{f \cdot C}\right) = V_P - N \cdot V_\Delta - \sum_{n=1}^{N} V^{\rho}_{n+1} + N \cdot V_\Phi - N \cdot \frac{I_o}{f \cdot C}$$

(4.3.14)

From equations (4.3.7) follows that:

$$V^{\rho}_n(T/2) = \rho \cdot \left(V_n(0) - V_{n-1}(0) - V^{\delta}_n\right)$$

(4.3.15)

Further, it is also known that:

Chargepump Modeling

$$V_n^\rho(T/2) = V_n(T/2) - V_{n-1}(T/2) - V_n^\delta$$

(4.3.16)

and:

$$V_n(0) - V_n(T/2) = \pm \frac{I_o}{f \cdot C}$$

(4.3.17)

For nodes 2 to N this leads to:

$$V_n^\rho = \frac{2 \cdot \rho}{1-\rho} \cdot \frac{I_o}{f \cdot C}$$

(4.3.18)

and at the input of the circuit:

$$V_{N+1}^\rho = \frac{\rho}{1-\rho} \cdot \frac{I_o}{f \cdot C}$$

(4.3.19)

At the start of the even phase, the output node voltage is equal to the voltage at node 1 minus V_Δ, minus V_1^ρ

$$V_1^\rho = \frac{I_o}{f \cdot C} \cdot \left(\frac{\rho}{1-\rho} + \frac{1}{2 \cdot \kappa} \cdot \left(\frac{\rho}{1-\rho} - \frac{1}{\ln(\rho)} \right) \right)$$

(4.3.20)

Then the voltage drops an additional $(I_o/2)/f\kappa C$ due to the load current I_o. This leads to the steady-state output voltage:

$$V_{Q_\rho}^\Omega = V_P - (N+1) \cdot V_\Delta + N \cdot V_\Phi - \left(N \cdot \frac{1+\rho}{1-\rho} + \frac{1}{2 \cdot \kappa} \cdot \left(\frac{1}{1-\rho} - \frac{1}{\ln(\rho)} \right) \right) \cdot I_o$$

(4.3.21)

The term $N(1+\rho)/(1-\rho)$ is equivalent to the expression for output resistance found by Van Steenwijk et al. [3,4]. However, this latter approximation is only valid for chargepumps with a large number of stages N.

The presence of series resistance in the diodes also results in a reduction of the output voltage ripple as illustrated in Figure 4.3.5. The amplitude of the voltage ripple lies within the region of:

$$\frac{I_o/2}{f \cdot C_o} \leq V_{Q_\rho}^\omega < \frac{I_o}{f \cdot C_o}$$

(4.3.22)

and depends on the value of ρ. A large value of ρ results in a lower voltage ripple which is favorable but also in a lower steady-state output voltage which is unfavorable. However, the lower steady-state output voltage can be largely avoided by adding some series resistance in the output diode D_1 only.

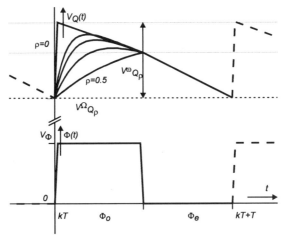

Figure 4.3.5 Output voltage waveform for different values of ρ.

As can be seen from Figure 4.3.4 and equation (4.3.21) the presence of series resistance, similar to leakage current, results in both a lower steady-state voltage and longer settling time. Realistic values for ρ are in the range of 10^{-2} to 10^{-4}. Consequently, the influence of series resistance can usually also be neglected.

4.3.3 Stray Capacitance

The nodes of the diode chain in a chargepump have parasitic capacitance to the substrate which influence the effective clock swing at those nodes. In circuit simulations these stray capacitances can be modeled as capacitors between the chargepump nodes and substrate as shown in Figure 4.3.6.

Transient Behavior

The effect of the stray capacitances is included in the general chargepump model by using a parameter σ defined as the ratio between the total stray capacitance C_n^σ connected to a node n and the pump capacitor C_n connected to that same node:

$$\sigma_n = \frac{C_n^\sigma}{C_n}$$

(4.3.23)

Chargepump Modeling

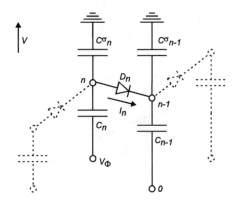

Figure 4.3.6 Stray capacitances

The stray capacitances appear as additional terms q^σ in the charge balance equations. For example, for the even phase:

$$q_{2m}(k)_{\Phi_e} + q^\sigma_{2m}(k)_{\Phi_e} + q_{2m-1}(k)_{\Phi_e} + q^\sigma_{2m-1}(k)_{\Phi_e}$$
$$= q_{2m}(k)_{\Phi_o} + q^\sigma_{2m}(k)_{\Phi_o} + q_{2m-1}(k)_{\Phi_o} + q^\sigma_{2m-1}(k)_{\Phi_o}$$

(4.3.24)

and,

$$\begin{cases} q_{2m}(k)_{\Phi_e} = C_{2m} \cdot \left(V_{2m-1}(k)_{\Phi_e} + V^\delta_{2m} - V_\Phi(k)\right) \\ q_{2m-1}(k)_{\Phi_e} = C_{2m-1} \cdot V_{2m-1}(k)_{\Phi_e} \\ q^\sigma_{2m}(k)_{\Phi_e} = \sigma_{2m} \cdot C_{2m} \cdot \left(V_{2m-1}(k)_{\Phi_e} + V^\delta_{2m}\right) \\ q^\sigma_{2m-1}(k)_{\Phi_e} = \sigma_{2m-1} \cdot C_{2m-1} \cdot V_{2m-1}(k)_{\Phi_e} \\ q_{2m}(k)_{\Phi_o} = C_{2m} \cdot V_{2m}(k)_{\Phi_o} \\ q_{2m-1}(k)_{\Phi_o} = C_{2m-1} \cdot \left(V_{2m-1}(k)_{\Phi_o} - V_\Phi(k)\right) \\ q^\sigma_{2m}(k)_{\Phi_o} = \sigma_{2m} \cdot C_{2m} \cdot V_{2m}(k)_{\Phi_o} \\ q^\sigma_{2m-1}(k)_{\Phi_o} = \sigma_{2m-1} \cdot C_{2m-1} \cdot V_{2m-1}(k)_{\Phi_o} \end{cases}$$

(4.3.25)

Further development is again similar to that of the parasitic free case and leads to an adapted system matrix A^σ and input matrix B^σ. The influence of the stray capacitance on the start-up behavior of a *3-stage* chargepump is shown in Figure 4.3.7. The bottom curve shows the predicted start-up together with a circuit simulation. In the inset it can be seen that model and simulation match accurately.

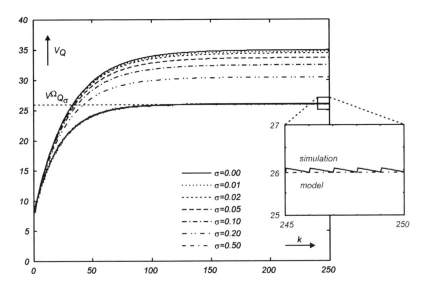

Figure 4.3.7 Influence of stray capacitance on start-up behavior

Steady-state Output Voltage

An expression for the steady-state output voltage $V^{\Omega}_{Q\sigma}$ can be obtained under the same assumptions as for the ideal steady-state expression, i.e. all pump capacitors have equal size $C_n=C$, the storage capacitor has size $C_o=\kappa C$ and all diodes have the same forward-bias voltage $V^{\delta}_n=V_\Delta$. Further, it is assumed that all diodes have the same stray capacitance parameter value $\sigma_n=\sigma$. If the clock driving capacitor C_n is low, the voltage on node n is equal to the preceding node $n+1$ minus V_Δ.

If the clock rises, the initial increase in the node voltage is equal to $V_\Phi C/(C+C^\sigma)=V_\Phi/(1+\sigma)$. Then an amount of charge equal to I_o/f is transferred to the succeeding capacitor C_{n-1} resulting in a voltage drop $I_o/f(C+C^\sigma)$. Consequently, the contribution of all stages equals:

$$\sum_{n=1}^{N}\left(V_\Phi \cdot \frac{1}{1+\sigma} - V_\Delta - \frac{I_o}{f \cdot C} \cdot \frac{1}{1+\sigma}\right) = \frac{N}{1+\sigma} \cdot V_\Phi - N \cdot V_\Delta - \frac{N}{1+\sigma} \cdot \frac{I_o}{f \cdot C}$$

(4.3.26)

Chargepump Modeling

At the start of the even phase, the output node voltage is equal to the voltage at node *1* minus V_Δ. Then the voltage drops an additional $(I_o/2)/f\kappa C$ due to the load current I_o. This leads to the steady-state output voltage:

$$V_{Q_\sigma}^\Omega = V_P - (N+1)\cdot V_\Delta + \frac{N}{1+\sigma}\cdot V_\Phi - \left(\frac{N}{1+\sigma} + \frac{1}{2\cdot\kappa}\right)\cdot I_o$$

(4.3.27)

This expression is similar to the steady-state expression published by Dickson [1] except that the latter gives the peak output voltage instead of the minimal output voltage. The output voltage ripple is not affected by stray capacitance.

As can be seen from Figure 4.3.7 and equation (4.3.27) the presence of stray capacitance, similar to parallel capacitance, results in a lower steady-state output voltage but also a shorter settling time. Again this shorter settling time cannot be exploited. Note that the relative insensitivity to stray capacitance compared to other voltage multipliers was the reason for choosing this chargepump configuration in the first place.

4.3.4 Parallel Capacitance

The charge transfer from one capacitor to the next through diodes is influenced by charge storage in the diodes. In circuit simulations the parallel capacitance can be modeled as a capacitor C^Φ_n in parallel with the diodes as shown in Figure 4.3.8.

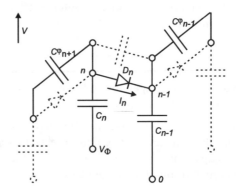

Figure 4.3.8 Parallel capacitances

The ratio between the parallel capacitance of diode D_n and pump capacitor C_n is defined as:

$$\varphi_n = \frac{C_n^{\varphi}}{C_n}$$

(4.3.28)

In case a *pn*-junction is used as a diode, two mechanisms cause charge storage: the *junction capacitance* and the *storage capacitance*.

The *junction* or *depletion capacitance* is associated with the charge dipole formed by the ionized donors and acceptors in the junction space charge region. Specifically, the junction capacitance relates the changes in charge at the edges of the depletion region to changes in the junction voltage. For an abrupt junction the junction capacitance per unit area is expressed by:

$$C_j = \sqrt{\frac{q\varepsilon_s}{2\cdot\left(\frac{1}{N_a}+\frac{1}{N_d}\right)\cdot(\Phi_i - V_a)}}$$

(4.3.29)

where, N_a and N_d are the doping densities in the p- and n-region, ϕ_i is the built-in junction potential and V_a is the applied diode bias voltage.

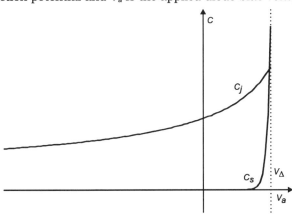

Figure 4.3.9 Junction and storage capacitance

The *storage* or *diffusion capacitance* is associated with the excess minority-carrier charge injected into the neutral regions under forward-bias conditions. This excess charge is proportional to the diode current and the minority lifetime. The storage capacitance per unit area is expressed by:

$$C_s = \frac{q}{kT}\cdot\left(J_{p0}\cdot\tau_p + J_{n0}\cdot\tau_n\right)$$

(4.3.30)

Chargepump Modeling

where, J_{po} and J_{no} are the injected minority-carrier current densities and τ_p and τ_n are the minority-carrier lifetimes in the quasi-neutral regions. Both C_j and C_s are plotted in Figure 4.3.9.

The (almost) exponential dependence on voltage of C_s causes it to be the dominant capacitance over most of the forward-bias range while in the reverse-bias range C_j is dominant.

In the chargepump circuit, the steep clock edges cause the charge transfer between subsequent capacitors to be almost instantaneous. Therefore, the current through the diodes is negligible at the end of each clock phase. Consequently, the minority-carrier storage is also negligible and only junction storage has to be taken into account. This is done as follows. The parallel capacitance C^φ is assumed to be independent of the applied junction voltage. A worst-case value for this capacitance is the junction capacitance at the forward-bias diode voltage V_Δ. A more realistic value is to take the mean of the junction capacitance value at V_Δ and the at maximum possible reverse-bias voltage which is $V_\Phi - V_\Delta$.

$$C^\varphi = \frac{1}{2} \cdot \left(C_j(V_\Delta) + C_j(V_\Delta - V_\Phi) \right)$$

(4.3.31)

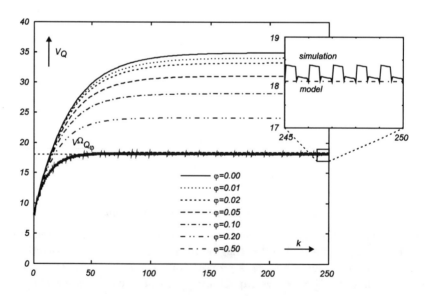

Figure 4.3.10 Influence of parallel capacitance on start-up behavior

Transient Behavior

The parallel capacitances appear as additional terms q^φ in the charge balance equations.

For example, in the even phase:

$$q_{2m}(k)_{\Phi_e} + q_{2m-1}(k)_{\Phi_e} + q^\varphi_{2m+1}(k)_{\Phi_e} + q^\varphi_{2m-1}(k)_{\Phi_e}$$
$$= q_{2m}(k)_{\Phi_o} + q_{2m-1}(k)_{\Phi_o} + q^\varphi_{2m+1}(k)_{\Phi_o} + q^\varphi_{2m-1}(k)_{\Phi_o}$$

(4.3.32)

However, this time four nodes instead of two nodes have to be taken into account when the charge balance is made.

$$\begin{cases} q_{2m}(k)_{\Phi_e} = C_{2m} \cdot \left(V_{2m-1}(k)_{\Phi_e} + V^\delta_{2m} - V_\Phi(k)\right) \\ q_{2m-1}(k)_{\Phi_e} = C_{2m-1} \cdot V_{2m-1}(k)_{\Phi_e} \\ q^\varphi_{2m+1}(k)_{\Phi_e} = \varphi_{2m+1} \cdot C_{2m+1} \cdot \left(V_{2m-1}(k)_{\Phi_e} + V^\delta_{2m} - V_{2m+1}(k)_{\Phi_e}\right) \\ q^\varphi_{2m-1}(k)_{\Phi_e} = \varphi_{2m-1} \cdot C_{2m-1} \cdot \left(V_{2m-1}(k)_{\Phi_e} - V_{2m-2}(k)_{\Phi_e}\right) \\ q_{2m}(k)_{\Phi_o} = C_{2m} \cdot V_{2m}(k)_{\Phi_o} \\ q_{2m-1}(k)_{\Phi_o} = C_{2m-1} \cdot \left(V_{2m-1}(k)_{\Phi_o} - V_\Phi(k)\right) \\ q^\varphi_{2m+1}(k)_{\Phi_o} = \varphi_{2m+1} \cdot C_{2m+1} \cdot \left(-V^\delta_{2m+1}\right) \\ q^\varphi_{2m-1}(k)_{\Phi_o} = \varphi_{2m-1} \cdot C_{2m-1} \cdot \left(V^\delta_{2m-1}\right) \end{cases}$$

(4.3.33)

From here on the model development is similar to that of the parasitic free case and leads to an adapted system matrix A^φ and input matrix B^φ.

The influence of parallel capacitance on the start-up behavior of a *3-stage* chargepump for different values of φ is shown in Figure 4.3.10. The bottom curve shows the predicted start-up together with a circuit simulation. In the inset it can be seen that model and simulation match accurately.

Steady-state Output Voltage

An expression for the steady-state output voltage including the effect of parallel capacitance is not so simple in this case. Due to the parallel capacitors, all nodes in the diode chain are coupled to each other all the time. Consequently, it is very hard to derive a closed analytical steady-state expression. For $N>2$ this expression already becomes very complex. Therefore, in this case an approximation for the steady-state output voltage $V^\varphi_{Q\varphi}$ is made which is valid under the same assumptions as for the

Chargepump Modeling

ideal steady-state expression, i.e. all pump capacitors have equal size $C_n = C$, the storage capacitor has size $C_o = \kappa C$ and all diodes have the same forward-bias voltage $V^s_n = V_\Delta$. Further, it is assumed that all diodes have the same parallel capacitance parameter value $\varphi_n = \varphi$. Then calculating:

$$\Delta V^\Omega = (I - A^\varphi)^{-1} \cdot B^\varphi \cdot u$$

(4.3.34)

yields and expression of the form:

$$\Delta V^\Omega_Q = \frac{1}{g(\varphi)} \cdot V_\Phi - \frac{1}{h(\varphi)} \cdot \frac{I_o}{f \cdot C}$$

(4.3.35)

An approximation of equation (4.3.35) can be made by making first order Taylor-approximations of $g(\varphi)$ and $h(\varphi)$. This leads to:

$$V^\Omega_{Q_\varphi} = V_P - (N+1) \cdot V_\Delta + \frac{N}{1 + \dfrac{1 + \kappa \cdot (4 \cdot N - 2)}{N \cdot \kappa} \cdot \varphi} \cdot V_\Phi - \frac{2 \cdot N \cdot \kappa + 1}{2 \cdot \kappa + \dfrac{2 \cdot (\kappa^2 \cdot (8 \cdot N - 4) + 3 \cdot \kappa + 1)}{2 \cdot N \cdot \kappa + 1} \cdot \varphi} \cdot I_o$$

(4.3.36)

Comparing the result of equations (4.3.34) and (4.3.36) for various parameter combinations showed that the approximation error is well within 1% of the steady-state output voltage $V^\Omega_{Q\varphi}$.

Due to crosstalk of the clock to the output, the output voltage ripple is higher than in the ideal case. An upper limit for this crosstalk ripple can be derived by considering only the parallel capacitance C^φ_1 of the output diode D_1. The rising edge of the clock driving C_1 will cause a rise of the output voltage equal to:

$$V^\omega_{Q_\varphi} = \frac{C_1}{C_1 + \dfrac{C^\varphi_1 \cdot C_o}{C^\varphi_1 + C_o}} \cdot \frac{C^\varphi_1}{C^\varphi_1 + C_o} \cdot V_\Phi = \frac{\varphi}{\kappa + \kappa \cdot \varphi + \varphi} \cdot V_\Phi$$

(4.3.37)

In Figure 4.3.11 the output voltage waveform with the effect of clock crosstalk is shown. The clock crosstalk ripple is added to the normal voltage ripple V^ω_Q.

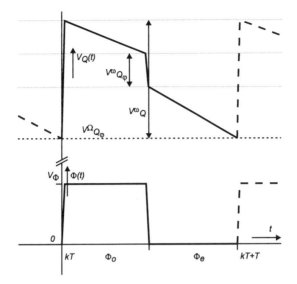

Figure 4.3.11 Output voltage waveform with clock crosstalk

As can be seen from Figure 4.3.10 and equation (4.3.36) the presence of parallel capacitance results in a lower steady-state voltage but also a shorter settling time. However, clearly, the shorter settling time cannot be exploited to improve chargepump performance.

4.3.5 Body-Effect

If MOSFETs are used to implement the diodes then the forward-bias voltage is about equal to the threshold voltage V_T of the transistors. The value of the threshold voltage increases when a reverse-biased voltage is applied to the source-substrate junction.

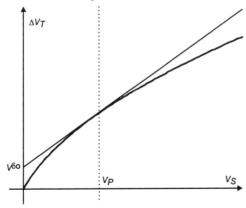

Figure 4.3.12 Threshold dependence on source voltage

Chargepump Modeling

This effect is often called the *body-effect* or *substrate-effect*. The dependence of the threshold voltage on the source-substrate voltage is expressed as:

$$V_T = V_{T0} + \Delta V_T = V_{T0} + \frac{\sqrt{2\varepsilon_s q N_a}}{C_{ox}} \cdot \left(\sqrt{2|\phi_p| + |V_{SB}|} - \sqrt{2|\phi_p|} \right)$$

(4.3.38)

where, V_{T0} is the threshold voltage if no substrate bias is applied, ϕ_p is the band bending in the substrate needed to invert the silicon surface. The dependence of the threshold voltage V_T on the source-substrate voltage V_{SB} can be approximated by a straight line as shown in Figure 4.3.12.

During normal operation, the lowest source voltage in an *N*-stage chargepump occurs at the node V_1 and is equal to V_P V_T. A fairly good approximation of the threshold voltage can be made by doing a first order Taylor-expansion around $V_s = V_P$. This results in:

$$V^\delta = \delta \cdot V_s + V^{\delta 0}$$

(4.3.39)

with:

$$\delta = \left. \frac{\partial V_T}{\partial V_s} \right|_{V_s = V_P}$$

(4.3.40)

Transient Behavior

In circuit simulations the body-effect can be modeled as a controlled voltage source in series with the diodes as shown in Figure 4.3.13.

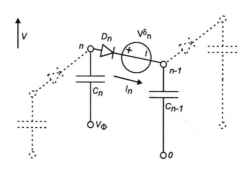

Figure 4.3.13 Body-effect

For this situation equation (4.3.39) can be written as:

$$V_n^\delta(k) = \delta_n \cdot V_{n-1}(k) + V_n^{\delta o}$$
$$= \delta_n \cdot \Delta V_{n-1}(k) + \delta \cdot V_{n-1}^\Gamma + V_n^{\delta o}$$
$$= \delta_n \cdot \Delta V_{n-1}(k) + V_n^\delta$$

(4.3.41)

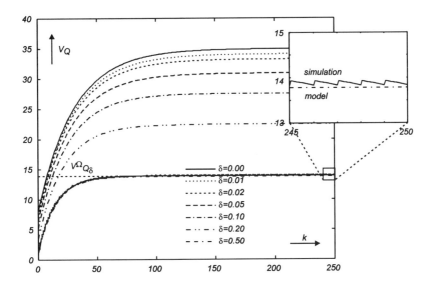

Figure 4.3.14 Influence of body-effect on start-up behavior

The charge balance remains unchanged since no charge is lost during transfer. In the even phase:

$$q_{2m}(k)_{\Phi_e} + q_{2m-1}(k)_{\Phi_e} = q_{2m}(k)_{\Phi_o} + q_{2m-1}(k)_{\Phi_o}$$

(4.3.42)

The influence of the body-effect can be included into the forward-bias voltage drop of the diode:

$$\begin{cases} q_{2m}(k)_{\Phi_e} = C_{2m} \cdot \left(V_{2m-1}(k)_{\Phi_e} + \delta_{2m} \cdot \Delta V_{2m-1}(k)_{\Phi_e} + V_{2m}^\delta - V_\Phi(k)\right) \\ q_{2m-1}(k)_{\Phi_e} = C_{2m-1} \cdot V_{2m-1}(k)_{\Phi_e} \\ q_{2m}(k)_{\Phi_o} = C_{2m} \cdot V_{2m}(k)_{\Phi_o} \\ q_{2m-1}(k)_{\Phi_o} = C_{2m-1} \cdot \left(V_{2m-1}(k)_{\Phi_o} - V_\Phi(k)\right) \end{cases}$$

(4.3.43)

Chargepump Modeling

The rest of the model development is similar to that of the parasitic free case and leads to an adapted system matrix A^δ and input matrix B^δ. The influence of the body-effect on the start-up behavior of a *3-stage* chargepump for different values of δ is shown in Figure 4.3.14. The bottom curve shows the predicted start-up together with a circuit simulation. In the inset it can be seen that model and simulation match accurately.

Steady-state Output Voltage

An expression for the steady-state output voltage $V^\Omega_{Q\delta}$ can be obtained under the same assumptions as for the ideal steady-state expression, i.e. all pump capacitors have equal size $C_n=C$, the storage capacitor has size $C_o=\kappa C$ and, at zero cathode voltage, all diodes have the same forward-bias voltage $V^{\delta o}_n=V_\Delta$. Further, it is assumed that all diodes have the same body-effect parameter value $\delta_n=\delta$. Due to the body-effect, the ground state of the chargepump is also changed. Equation (4.3.41) leads to:

$$V_{n-1} = V_n - V_n^\delta = \frac{V_n}{1+\delta} - \frac{V_\Delta}{1+\delta}$$

(4.3.44)

Consequently, the ground state is given by:

$$V^\Gamma_{Q\delta} = \frac{V_P}{(1+\delta)^{N+1}} - \sum_{n=1}^{N+1} \frac{V_P}{(1+\delta)^n} = V_P \cdot (1+\delta)^{-(N+1)} - V_\Delta \cdot \frac{1-(1+\delta)^{-(N+1)}}{\delta}$$

(4.3.45)

If the clock driving capacitor C_n is low, the voltage on node n is equal to the preceding node minus V^δ_n. If the clock rises, the initial increase in the node voltage is equal to V_Φ. Then an amount of charge equal to I_o/f is transferred to the succeeding capacitor C_{n-1} resulting in a voltage drop I_o/fC. Similar to the contribution of the supply voltage V_P in equation (4.3.45) the contribution of the pumping stages then equals:

$$\sum_{n=1}^{N} \frac{V_\Phi - \frac{I_o}{f \cdot C}}{(1+\delta)^n} = \frac{1-(1+\delta)^{-N}}{\delta} \cdot \left(V_\Phi - \frac{I_o}{f \cdot C}\right)$$

(4.3.46)

At the start of the even phase, the output node voltage is equal to the voltage at node *1* minus V_Δ. Then the voltage drops an additional $(I_o/2)/f\kappa C$ due to the load current I_o. This leads to the steady-state output voltage:

Modeling of Parasitic Effects

$$V_{Q_\delta}^\Omega = V_P \cdot (1+\delta)^{-(N+1)} - V_\Delta \cdot \frac{1-(1+\delta)^{-(N+1)}}{\delta} - V_\Phi \cdot \frac{1-(1+\delta)^{-N}}{\delta} - \left(\frac{1-(1+\delta)^{-N}}{\delta} + \frac{1}{2\cdot\kappa} \right) \cdot l_O$$

(4.3.47)

This expression in similar to the one published by Witters *et al.* [2]. However, in expression (4.3.47) the influence of the storage capacitor is included while, on the other hand, Witters *et al.* included the effect of stray capacitance. The output voltage ripple is not influenced by the body-effect.

As can be seen from Figure 4.3.14 and equation (4.3.47) body-effect results in a lower steady-state output voltage and a shorter settling time. Again there is no point in exploiting the shorter start-up time. Realistic values for the body-effect parameter δ are in the range of 0.05.

4.3.6 Combination of Parasitic Effects

In the previous sections the most important parasitic effects that influence the behavior of chargepump circuits were treated separately. However, usually a combination of effects occurs.

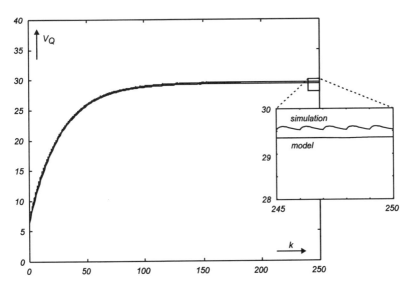

Figure 4.3.15 Start-up behavior with $\lambda=0.01$, $\rho=0.01$, $\sigma=0.02$, $\varphi=0.01$, $\delta=0.05$

In this case a worst-case approximation can be used to predict the overall chargepump behavior. This can be done as follows. First, the effect of all

Chargepump Modeling

parasitics are calculated separately. Then the difference of each calculation compared to the ideal case is calculated. These differences are then summed together and subtracted from the ideal case. This will give a worst-case estimation as long as it can be assumed that the parasitic effects do not enhance eachother. An example is shown in Figure 4.3.15. The parameters used in this calculation are: $f=1.0MHz$, $I_o=1.0mA$, $C=1nF$, $\kappa=9$, $V_P=V_\phi=10V$ and $V_\Delta=0.5$ and for the parasitics: $\delta=0.05$, $\sigma=0.02$, $\varphi=0.01$, $\rho=0.01$ and $\lambda=0.01$. As can be seen the model and simulation match quite well both for the dynamic and static part. The model predicts a steady-state output value which is a bit lower than the value obtained from the simulation. This yields a practical worst-case value which can be calculated much faster than by doing circuit simulations.

4.4 Conversion Efficiency

The power consumption and conversion efficiency of chargepump circuits are especially important in applications in which the chargepump is used as a voltage power supply.

4.4.1 Power Consumption and Efficiency

The transient power consumption can be calculated by calculating the amount of charge $Q(k)$ that is injected into the circuit each clock cycle k. All charge is injected through the input node and the clock drivers. It is reasonable to assume that both the input node and the clock drivers are connected to the same supply voltage V_P as indicated in Figure 4.4.16.

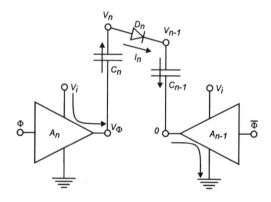

Figure 4.4.16 Charge transfer

Therefore the total amount of power consumed in one clock cycle is given by:

$$P_P(k) = V_P \cdot f \cdot \sum_{n=1}^{N+1} Q_n(k)$$

(4.4.48) In the even phase, charge is injected through the capacitors connected to inverted clock $\overline{\Phi}$, that is, the capacitors with an even number 2m. The amount of charge injected at each node is found to be:

$$Q_{2m}(k) = -C_{2m} \cdot \left(\Delta V_{2m}(k)_{\Phi_e} - \Delta V_{2m}(k-1)_{\Phi_o} - V_\Phi(k) \right)$$

(4.4.49)

For the odd phase a similar calculation can be done. In steady-state, the amount of charge transferred during each transition is equal to $Q = I_o/f$. Therefore, an expression for the steady-state input power of a chargepump circuit is given by:

$$P_P^\Omega = V_P \cdot f \cdot \sum_{n=1}^{N+1} I_o/f = (N+1) \cdot V_P \cdot I_o$$

(4.4.50)

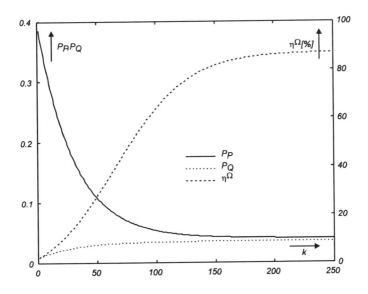

Figure 4.4.17 Power consumption and conversion efficiency during start-up.

Conversion Efficiency

The conversion efficiency $\eta(k)$ of a chargepump is defined as:

Chargepump Modeling

$$\eta(k) = \frac{P_Q(k)}{P_P(k)}$$

(4.4.51)

where $P_Q(k)$ is the available output power which is simply the product of the output voltage $V_Q(k)$ and load current $I_o(k)$. In steady-state the efficiency η^Ω converges to:

$$\eta^\Omega = \frac{V_Q^\Omega}{(N+1) \cdot V_i}$$

(4.4.52)

In Figure 4.4.17 the input power, output power and conversion efficiency during start-up of a *3-stage* chargepump are shown. The other parameters used in this calculation are: $f=1.0MHz$, $I_o=1.0mA$, $C=1nF$, $\kappa=9$, $V_P=V_\Phi=10V$ and $V_A=0.5V$. As can be seen, the input power during start-up is much higher than in steady-state, which results in a poor efficiency. Consequently, it can be advantageous to avoid frequent on/off switching of the chargepump.

4.4.2 Influence of Parasitics

The parasitic effects described in this section influence the power consumption and conversion efficiency of the chargepump.

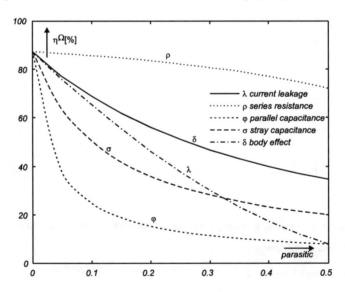

Figure 4.4.18 *Conversion efficiency versus parasitics.*

Conversion Efficiency

Calculation of the input and output power is similar to the ideal case, that is, the charge injected into the circuit after each clock transition can be calculated. The influence of the parasitics on the steady-state conversion efficiency is shown in Figure 4.4.18.

As can be seen the effect of current leakage, stray capacitance and parallel capacitance is most severe. In the case series resistance or body-effect is present, basically only the value of the output voltage is affected. In steady-state, the input power is the same as in the ideal case. However, if current leakage, stray capacitance or parallel capacitance are present, extra charge has to be injected to compensate for the leaked charge and charging and discharging of the parasitic capacitances. Therefore, the input power increases.

A parasitic effect which has not been considered before is the capacitance between the bottom plate of the pump capacitors and substrate. This parasitic does not influence the chargepump behavior but it does represent an extra load for the clock drivers as shown in Figure 4.4.19.

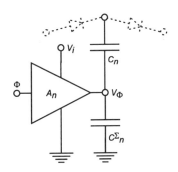

Figure 4.4.19 Substrate capacitance.

The substrate capacitances are charged and discharged each clock cycle and can have a rather high value compared to the pump capacitor value itself. After all, this was the reason for using the Dickson configuration for integration of voltage multipliers in the first place. The substrate capacitances are usually nonlinear but can be approximated with a linear value C^Σ. The additional power consumption per clock driver is given by:

$$P_n^\Sigma(k) = V_P \cdot V_\Phi(k) \cdot C_n^\Sigma \cdot f$$

(4.4.53)

The influence of the substrate capacitance has to be taken account in the dimensioning of the clock drivers since additional current has to be supplied to charge and discharge of the substrate capacitances.

4.5 Conclusion

An extended chargepump model has been presented that can be used to describe both the dynamic and steady-state behavior of chargepump circuits with an arbitrary number of stages. The model is based upon analysis of a charge balance between adjacent capacitors in the chargepump.

A number of previously published chargepump models have been presented. All these models have been shown to derive from the new extended model by choosing the appropriate parameters.

The effect of five parasitic effects are included in the model. These are, current leakage, series resistance, parallel capacitance ,stray capacitance and body-effect. The inclusion of these effects is done with simple modifications in the diode modeling.

The effect of the parasitics on the conversion efficiency has been analyzed. It has been demonstrated that especially parasitic capacitances deteriorate the conversion efficiency.

The validity of the model has been demonstrated by comparison with circuit simulations.

The charge balance analysis can easily be extended to describe the dynamic and steady-state behavior of switched capacitor step down converters that were introduced in Appendix 3.8.

4.6 References

[1] Dickson, J.F. "On-Chip High-Voltage Generation in MNOS Integrated Circuits Using an Improved Voltage Multiplier Technique", IEEE Journal of Solid-State Circuits, Vol.11, No.3, pp.374-378, June. 1976.
[2] Witters, J.S., G. Groeseneken, H.E. Maes, "Analysis and Modeling of On-Chip High-Voltage Generator Circuits for Use in EEPROM Circuits", IEEE Journal of Solid-State Circuits, Vol.24, No.5, pp.1372-1380, Oct. 1989.
[3] Steenwijk, G. van, K. Hoen, H. Wallinga, "Analysis and Design of a Charge Pump Circuit for High Output Current Applications", ESSCIRC'93, pp.118-121, Sep. 1993.
[4] Steenwijk, G. van, "Analog Applications of the VIPMOS EEPROM", Ph.D. Thesis University of Twente, 1994, ISBN 90-9007200-4.
[5] Cataldo, G. Di, G. Palumbo, "Double and Triple Charge Pump for Power IC: Ideal Dynamical Models to an Optimised Design", IEE Proceedings-G, Vol.140, No.1, pp.33-38., Feb 1993.
[6] Cataldo, G. Di, G. Palumbo, "Double and Triple Charge Pump for Power IC: Dynamic Models Which Take Parasitic Effects into Account", IEEE Transactions on Circuits and Systems - I, Vol.40, No.2, pp.92-101, Feb. 1993.

5

BCD Audio Amplifiers

In this chapter an integrated *100W* audio power amplifier is presented realized in a *BCD* technology. Several different amplifier topologies are discussed and compared on their suitability for integration in a *BCD* technology. Based on this comparison one particular amplifier topology is selected and developed further. A detailed description of the design of this amplifier is presented in which a chargepump circuit is used in order to achieve rail-to-rail output capability.

5.1 Introduction

As explained in Chapter 1 the output stage of a single-ended class *AB* power amplifier contains two power transistors operating together in push-pull. In most *BCD* technologies only *n-type DMOS* transistors are available as power transistors. Consequently, the output stage of a *BCD* power amplifier will have a totempole structure as shown in Figure 5.1.1. The design of the rest of the amplifier is now dedicated to generating the appropriate signals V_H and V_L to drive the gates of the output transistors with. In the design of audio amplifiers three subjects play an important role.

BCD Audio Amplifiers

First, the input signal V_i has to be amplified with as little *distortion* as possible.

Second, the *output resistance* of the amplifier has to be low in order to obtain good loudspeaker damping.

Third, in order to minimize *dissipation* in the output transistors the output stage operates in class *AB*. In class *AB* operation only one of output transistors supplies current at a time to the load while the other output transistor is turned off or, preferably, conducts a small *residual current*. The *quiescent current* I_q in the output stage of a modern integrated audio amplifier is in the order of *10mA* whereas the maximum output current I_o can be as high as *10A*.

Due to this extreme ratio between output and quiescent currents, it is very difficult to achieve low distortion levels.

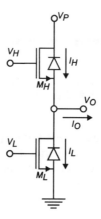

Figure 5.1.1 DMOS totempole structure

In the following paragraphs the various sources of distortion are discussed in detail and two alternative amplifier types with low output resistance are presented and compared. Finally, some design considerations with respect to circuit design in *BCD* technology are given.

5.1.1 Sources of Distortion

Distortion in class *AB* amplifiers basically has two causes. In the first place, the inherent nonlinearity of the transistors always causes a certain amount of distortion. Distortion caused by device nonlinearity is called *large signal distortion*. Second, the nature of class *AB* operation is inherently nonlinear. In Figure 5.1.2 the currents I_H and I_L through the

Introduction

output transistors are shown together with the output current I_o which is the difference between I_H and I_L. Although I_o is a perfect sinewave, the currents I_H and I_L are clearly highly distorted with respect to I_o. So the class AB operation introduces distorted signals inside the amplifier because of the nonlinear splitting of the input signal.

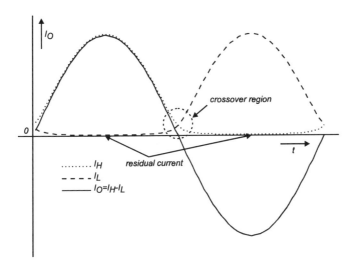

Figure 5.1.2 Nonlinear currents in a class AB output stage

A type of distortion which is typical for class AB operation occurs in the crossover region indicated in Figure 5.1.2 and is called *crossover distortion*. A detailed description of the various sources of *large signal* and *crossover* distortion is given in the following sections.

Large Signal Distortion

Large signal distortion is caused by the nonlinearity of the transfer characteristics of the output transistors. Although this source of distortion is present in all electronic circuits it is more significant in class AB operation, since the output transistors are operated from maximally-on to near switch-off and from the saturation region deep into the triode region. A distinction can be made in *static* and *dynamic* large signal distortion.

Static large signal distortion is generated if a nonlinear transistor is in the signal path. Consider, for example, the circuit shown in Figure 5.1.3(a). Because of the nonlinear transfer characteristic of the DMOS transistor M_o the output current I_o will contain harmonic components at multiples of the frequency of the input signal. The result of this nonlinearity is often called *harmonic distortion*. Static large signal distortion can be compensated for to a certain extent by making the inverse of the transistor nonlinearity with a replica transistor. For the circuit in Figure 5.1.3(a) this results in the well-known current mirror shown in Figure 5.1.3(b). The input current I_i is now transformed into a nonlinear voltage V_i that in turn is transformed into a linear output current I_o.

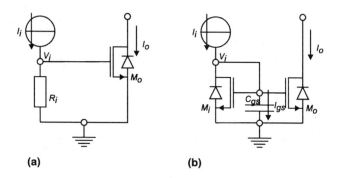

(a)   (b)

Figure 5.1.3 Large signal distortion (a) static (b) dynamic

For low frequencies this compensation works very good. However, for higher frequencies, the influence of the gate-source capacitances C_{gs} of the DMOS transistors becomes significant. The nonlinear voltage V_i causes a nonlinear current I_{gs} through this capacitor which is subtracted from the input current I_i thus causing so-called *dynamic large signal distortion*. Since this distortion is caused by reactive components it is also known as *reactive harmonic distortion* [1].

Both static and dynamic large signal distortion are significant sources of distortion in a DMOS class AB output stage. In many output stage topologies it is not possible to use a compensation method such as in the current mirror. If compensation is possible then its effectiveness depends on the matching of the transistors. Since power transistors in output stages are very large, it is not realistic to use an equally sized transistor for compensation. The ratio between output and compensation transistor is in general quite large (typically *1:100* to *1:1000*) which severely limits the matching whereas differences in temperature and drain-source voltage cause further mismatch. The gate-source capacitance of the output transistors, which is typically in the order of a few hundred picoFarads, can cause significant dynamic large signal distortion at relatively low frequencies.

Crossover Distortion

Crossover distortion results from the push-pull operation of the output transistors. This distortion appears when the output current of the output stage changes direction. In this case, the control of the output voltage is transferred from on output transistor to the other. Similar to large signal

Introduction

distortion a distinction can be made in *static* and *dynamic* crossover distortion.

Static crossover distortion is caused by variations in gain and output resistance of the output stage in the crossover region. These variations result from the nonconjugate nature of the output transfer characteristic as shown in Figure 5.1.4 and is essentially frequency independent. In pure class B operation static crossover distortion can easily be understood since in the crossover region both output transistors are turned off simultaneously resulting in a so-called "*deadband*". This causes the gain to drop and output resistance to rise rapidly. This deadband is eliminated by biasing the output transistors in such a way that they conduct simultaneously in the crossover region. However, this may lead to so-called "g_m-*doubling*" [5] which causes gain to increase and output resistance to drop in the crossover region.

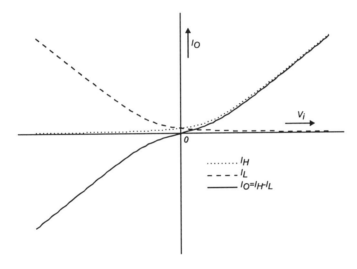

Figure 5.1.4 Transfer characteristic of class B and class AB stages

Dynamic crossover distortion can be caused by time delay during crossover. In many class AB stages, one output transistor is conducting current while the other is turned off. During crossover the conducting transistor is turned off while the other transistor is turned on. If the timing during crossover is not accurate this results in distortion of the output signal. Differences in timing result from the large charge and discharge currents needed to drive the output transistors. In output stages with bipolar power transistors, the turn-off speed is limited by minority carrier storage in the base region resulting in so-called *switching distortion*. If DMOS power transistors are used, the large gate-source capacitance limits the switching speed. Switching distortion can be reduced significantly if the inactive transistor is not turned off completely but conducts a small *residual current* [11].

Crossover distortion is basically a complex form of large signal distortion. However, there are some differences in the manifestation of both types of distortion.

The distortion spectrum of large signal distortion usually contains predominantly low-order harmonics. The level of this distortion tends to increase with increasing signal level.

Crossover distortion appears as sharp peaks in the output signal in the crossover region. Consequently, the distortion spectrum of crossover distortion contains many high frequency components. The level of this distortion tends to decrease with increasing signal level.

5.1.2 Output Resistance

One of the requirements for audio power amplifiers is a low output resistance which is needed for good loudspeaker damping. In order to satisfy this requirement most audio amplifiers are implemented as a high gain voltage amplifier with global feedback as shown in Figure 5.1.5(a). The closed loop voltage gain A'_v of this configuration is given by:

$$A'_v = \frac{A_v}{1 + A_v \cdot \beta} \approx \frac{1}{\beta}$$

(5.1.1)

where β is the *feedback factor* determined by the resistors R_a and R_b:

$$\beta = \frac{R_b}{R_a + R_b}$$

(5.1.2)

The current through resistors R_a and R_b is assumed to be negligible. The open loop output resistance R_o of the voltage amplifier is already low. The closed loop output resistance R'_o is reduced further by the loop gain $A_v\beta$:

$$R'_o = \frac{R_o}{1 + A_v \cdot \beta} \approx \frac{R_o}{A_v \cdot \beta}$$

(5.1.3)

The voltage amplifier usually has a open loop voltage gain A_v of *80dB* with a dominant pole at about *1kHz* while the closed loop gain A'_v lies between *26dB* and *30dB*.

Introduction

An alternative approach is to use a transconductance amplifier instead of a voltage amplifier as shown in Figure 5.1.5(b). The closed loop voltage gain A'_v for this configuration is given by:

$$A'_v = \frac{G_m \cdot Z_L}{1 + G_m \cdot Z_L \cdot \beta} \approx \frac{1}{\beta}$$

(5.1.4)

The open loop output resistance R_o of the transconductance amplifier is very high. However, the voltage feedback again reduces the closed loop output resistance R'_o:

$$R'_o = \frac{1}{\frac{1}{R_o} + G_m \cdot \beta} \approx \frac{1}{G_m \cdot \beta}$$

(5.1.5)

Consequently, a low closed loop output resistance can be realized when the transconductance G_m is sufficiently large. A typical value for G_m is *1000A/V* while the closed loop gain $1/\beta$ lies between *26dB* and *30dB* resulting in a closed loop output resistance R'_o between *10mΩ* and *30mΩ*.

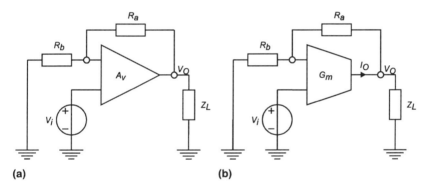

Figure 5.1.5 Feedback amplifiers (a) voltage amplifier (b) transconductance amplifier

Voltage amplifiers are typically composed of three stages as shown in Figure 5.1.6. The input stage G_1 is a differential voltage to single-ended current converter. The resistor R_1 represents the output resistance of the input stage G_1 in parallel to the input resistance of the second stage G_2. The second stage G_2 is an inverting voltage to current converter that, loaded with R_2, provides additional voltage gain. The resistor R_2 represents the output resistance of the second stage G_2 in parallel to the input resistance of the third stage A_3. The third stage A_3 is a low gain or unity-gain voltage buffer that separates the gain stages from the load.

Frequency compensation is often done by connecting a Miller capacitor C_m between the input and output of the inverting gain stage G_2. The open loop voltage gain A_v is given by:

$$A_v = \frac{G_1 \cdot R_1 \cdot G_2 \cdot R_2 \cdot \left(1 - j\omega \cdot \frac{C_m}{G_2}\right)}{1 + j\omega \cdot (R_1 + R_2 + R_1 \cdot R_2 \cdot G_2) \cdot C_m}$$

(5.1.6)

Although it is also possible to include the unity-gain buffer in the Miller loop which helps to reduce distortion of the buffer [12] this is usually not done since it can cause instability with complex loads.

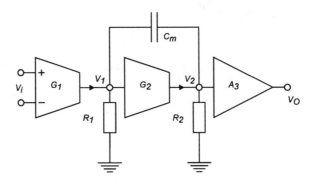

Figure 5.1.6 Voltage amplifier structure

A typical transconductance amplifier is shown in Figure 5.1.7. The input stage G_1 is similar to that of a voltage amplifier. However, in this case the output stage is a voltage to current converter G_2.

Frequency compensation is done by introducing a dominant pole at the internal node V_1. The open loop transconductance G_m is given by:

$$G_m = \frac{G_1 \cdot R_1 \cdot G_2}{1 + j\omega \cdot R_1 \cdot C_1}$$

(5.1.7)

The open loop voltage gain A_v of the output stage is determined by the open loop transconductance G_m but also by the load impedance Z_L:

$$A_v = G_m \cdot Z_L$$

(5.1.8)

Frequency compensation can also be done by connecting a Miller capacitor across the (inverting) output stage G_2 as indicated with the dashed line in

Figure 5.1.7. However, since the voltage gain of this stage is also determined by the load impedance Z_L it is very difficult to ensure the stability of the Miller loop for all possible load impedances. On the other hand, without Miller compensation the amplifier is not stable if no load is attached to the output. If open load stability is required a Miller capacitor is unavoidable.

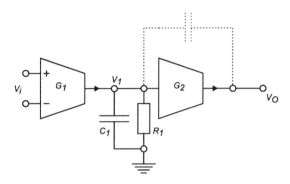

Figure 5.1.7 Transconductance amplifier structure

In both amplifier types, the distortion of the complete amplifier is dominated by the distortion in the output stage [2-9]. Distortion is reduced by applying negative feedback. However the amount of negative feedback that can safely be applied is limited for stability reasons. Therefore, much design effort is put in making the output stage of power amplifiers as linear as possible. In Section 5.2 some alternative output stage topologies are presented.

5.1.3 Circuit Design in BCD Technology

A description of the *BCD* technology used for the designs presented in this book is given in Chapter 2 together with a description of the available components. The logical choice for the output power transistors is the *VDMOS* transistor. For the remaining circuits a choice has to be made between *LDMOS*, *EPMOS*, *CMOS*, vertical *NPN* and lateral *PNP* transistors. Before starting the design of circuits, a number of assumptions is made.

From the available transistors only the *VDMOS*, *LDMOS* and *EPMOS* transistors are capable of handling the supply voltage (*60V*). Consequently, the use of these transistors is inevitable on several places in the circuit. The characteristics of the *LDMOS* transistors are quite attractive. It has a relatively high gain factor due to the short channel and a very high output impedance due to the drift region. The *EPMOS* transistors on the other hand have inferior characteristics. The rather provisional structure of the drift region of the transistor causes it to have a relatively low gain factor and strongly bias dependent output impedance. Consequently, in the signal path the use of *LDMOS* transistors is preferred.

All subcircuits in the amplifier are assumed to operate in class *A*. This means that the signal current is always smaller than the biasing current. It is expected that this makes the design of stable local feedback loops less difficult.

Finally, it is assumed that an on-chip chargepump is available which is capable of supplying a current of a few milliAmperes at an output voltage V_Q of 70V. This chargepump output voltage can be used to drive the high-side output transistor. As is explained in Chapter 3, it is preferred that the chargepump is loaded with a constant current in order to minimize the voltage ripple and simplify output voltage regulation.

In the following sections the design of a *BCD* amplifier is presented. First, in section 5.2 a high level comparison is made between different output stage topologies after which one stage is selected for further development. The considerations given in this section are general in nature and are not restricted to a specific *BCD* technology. Next, in section 5.3 a description of the circuits and techniques used in the implementation of the selected amplifier topology in a specific *BCD* technology are presented. Because one specific *BCD* technology is used, many of the considerations in this section are related to this technology.

5.2 Output Stage Topologies

As was explained in the previous section, two types of output stage exist. The first, and most commonly used type is a voltage buffer. A voltage buffer has a low open loop output impedance and the voltage gain is usually near unity. If complementary output transistors are available the simplest voltage buffer implementation uses the *common drain* configuration, or *source follower*, shown in Figure 5.2.1(a).

The second type is a transconductance stage. The open loop output impedance is fairly high. If complementary output transistors are available the simplest implementation uses the *common source* configuration shown in Figure 5.2.1(b).

However, in most *BCD* technologies only *n-type DMOS* transistors are available which means that neither of the basic complementary topologies can be used unchanged. In order to make an all *n-DMOS* output stage that behaves like a common drain stage, an implementation has to be found that makes the low-side transistor M_L behave as a *p-type DMOS* transistor. Conversely, in order to make an all *n-DMOS* output stage that behaves like a common source stage, an implementation has to be found that makes the high-side transistor M_H behave as a *p-type DMOS* transistor.

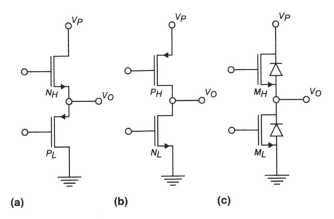

Figure 5.2.1 *Output stage configurations (a) complementary common drain (b) complementary common source (c) all n-DMOS*

In the following sections some implementations of both *common drain* and *common source* stages are presented and compared with respect to *distortion* and *quiescent current control*. Following this comparison, one stage is selected for further development.

5.2.1 Common Drain Stages

Most voltage buffer output stages are based on the complementary bipolar common-collector circuit shown in Figure 5.2.2(a). Quiescent current control is performed by the translinear loop formed by the transistors Q_1, Q_2, Q_H and Q_L which causes the currents I_H and I_L to obey the well-known geometric mean law:

$$I_H \cdot I_L = I_q^2$$

(5.2.9)

In the ideal case both output transistors conduct current at all time. The circuit has a low output impedance and the voltage gain is slightly less than unity.

A similar output stage using only *NPN* power transistors is shown in Figure 5.2.2(b). In this stage the transistors Q_3 and Q_L form a so-called *composite-PNP*. Due to the local feedback loop formed by Q_3 and Q_L the combination behaves as a power *PNP* transistor. Therefore, this stage is usually called a *quasi-complementary* stage. A drawback of this composite-PNP is that it tends to be unstable at high current levels and complex loads [11].

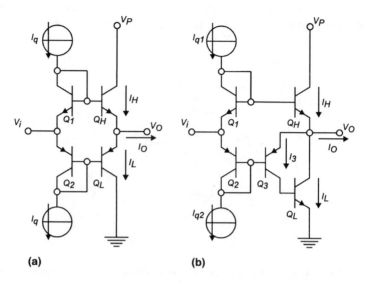

Figure 5.2.2 Bipolar voltage buffers (a) complementary (b) quasi-complementary

Quasi-complementary DMOS Voltage Buffer

The bipolar quasi-complementary stage can easily be translated to the all n-DMOS equivalent shown in Figure 5.2.3 which was published by Brasca and Botti [13]. In this stage a differential amplifier A_l is used to make a *composite p-DMOS* with a local feedback loop. In the quiescent point, the output voltage V_o will be about equal to the input voltage V_i as long as the gain of the amplifier A_l is higher than unity. In this case the DMOS transistors M_h and M_H have the same gate-source voltage. Consequently the quiescent current in the output transistors is given by:

$$I_H = I_L = N \cdot I_q$$

(5.2.10)

where N is the ratio between the gate widths of M_H and M_h. Note that the current source I_q has to be connected to a voltage V_Q that is higher than the supply voltage V_P which is a necessity for rail-to-rail output capability.

An advantage of this stage is that it has low complexity while it still provides accurate quiescent current control. However, there are also some drawbacks.

During negative excursions of the output signal the voltage difference between V_o and V_i is very small as long as the amplifier A has enough gain. Consequently, the current through M_H barely changes and stays almost equal to the quiescent current NI_q. However, during positive excursions, the gate-source voltage V_{gs} of M_H varies substantially when it needs to source current to the load. This results in two undesirable effects.

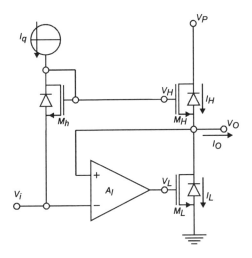

Figure 5.2.3 Quasi-complementary DMOS voltage buffer

First, the nonlinear V_{gs} variations of transistor M_H cause considerable *large signal distortion*. This distortion is higher in DMOS output stages compared to bipolar output stages because the transconductance of bipolar transistors is higher.

Second, the voltage difference between V_o and V_i causes the low-side transistor M_L to be turned off completely which may result in dynamic crossover distortion. In order to prevent the latter a clamp on the gate of M_L is used to keep M_L conducting.

It is remarkable that the behavior of this stage is very different for positive and negative output voltage excursions. This is mainly caused by the highly asymmetrical topology of this stage. Due to the asymmetrical nature it can be expected that the harmonic distortion contains a high level of even harmonics.

In order to improve the symmetry of the stage, three approaches can be used. The first is to make the low-side behave more like the high-side. The second is to make the high-side behave more like the low-side. The third is to use something completely different.

Improved Quasi-complementary DMOS Voltage Buffer

An attempt at the first approach to obtain a more symmetrical, quasi-complementary stage is shown in Figure 5.2.4. In this stage a combination of two identical transconductors G_{lx} and G_{ly} is used to copy the voltage difference between V_l and V_o to a voltage difference between gate and source of M_L. In this way the DMOS transistors M_h, M_l, M_H and M_L behave as if they are connected in a MOS equivalent of a translinear loop [14].

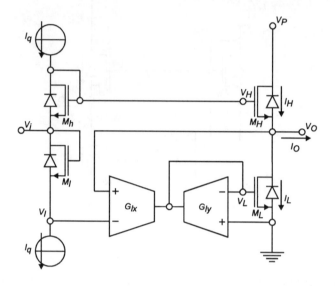

Figure 5.2.4 *Quasi-complementary DMOS voltage buffer with improved symmetry*

If it is assumed that the DMOS transistors have a square law transfer characteristic and are all biased in the saturation region, the translinear loop formed by the four DMOS transistors causes the currents I_H and I_L to obey:

$$\sqrt{I_H} + \sqrt{I_L} = 2 \cdot \sqrt{N \cdot I_q}$$

(5.2.11)

where N is the ratio between the gate widths of M_h/M_H and M_l/M_L respectively. The quiescent value of I_L and I_H can readily be found to be equal to NI_q and as long as the voltage copier G_{lx}/G_{ly} operates correctly, the low-side and high-side driver behave symmetrically. However, relation (5.2.11) becomes invalid as soon as either I_L or I_H exceeds the value of $4I_q$. In this case the opposite output transistor is turned off. The transistor currents I_H and I_L as a function of the output current I_o are shown in Figure 5.2.5. For comparison the currents corresponding to a *geometric mean law* are also shown.

In this stage a clamp for the high-side transistor M_H as well as the low-side transistor M_L is needed to prevent turn-off. Further, during negative excursions of the output signal, the large signal distortion is now equal to that during positive excursions.

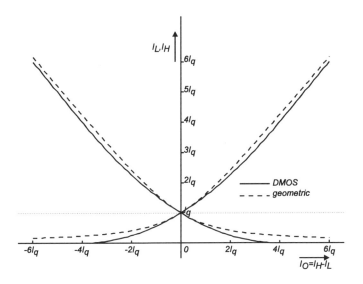

Figure 5.2.5 Output transistor currents as a function of output current

Another Improved Quasi-complementary DMOS Voltage Buffer

An attempt at the second approach to obtain a more symmetrical, quasi-complementary stage is shown in Figure 5.2.6. This output stage was originally used by Lish [15] for a power OpAmp in a *BiCMOS* technology who named it a *floating buffer stage* referring to the floating amplifier A_h. In this stage the symbols indicated by Σ_1 and Σ_2 are voltage summers.

In the quiescent point the output voltage V_o is nearly equal to the input voltage V_i. In this case the voltage summers Σ_h and Σ_l will force the gate-source voltages of M_H and M_L equal to the gate-source voltages of M_h and M_l respectively. Consequently, the quiescent current in M_H and M_L is, similar to the previous stage, determined by the current I_q and the ratio N between the gate widths of M_H/M_h and M_L/M_l. Because this stage has an amplifier in the high-side driver as well, the large signal distortion caused by nonlinear V_{gs} variations of M_H is reduced substantially in this stage compared to the previous stages.

Special attention has to be paid to the voltage summers Σ_h and Σ_l. If they are implemented as linear voltage summers the output transistors M_H and M_L have the same turn-off problem as described previously.

To avoid this, the summers are implemented in such a way that only positive voltages from amplifiers A_h and A_l are added to the quiescent voltages determined by M_h and M_l. Consequently, the combination of Σ_h

BCD Audio Amplifiers

with M_h and Σ_l with M_l operate as voltage clamps for the outputs of A_h and A_l.

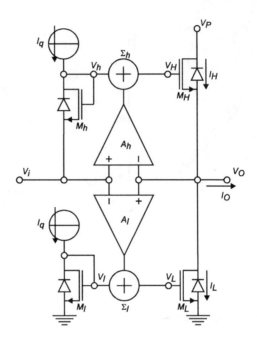

Figure 5.2.6 Quasi-complementary DMOS voltage buffer with improved symmetry

Minimum Selector Voltage Amplifier

The previous stages have in common that they require a voltage clamp to prevent the turn-off of one or both output transistor. In bipolar stages, turn-off of the output transistors also occurs because the translinear loop is disturbed by the resistance in the emitters of Q_H and Q_L.

A solution that avoids turn-off can be found in the circuit by Seevinck et al [11]. This circuit was originally intended for a power amplifier with bipolar output transistors. In Figure 5.2.7 the same circuit with DMOS output transistors is shown. The operation of the circuit is as follows. The output transistors M_L and M_H are driven by a fully differential amplifier formed by differential pair Q_1 and Q_2 and the two controlled current sources I_c which conduct the same current. The differential output level is controlled by the input voltage V_i while the common level is controlled by the two current sources I_1.

The replica transistors M_h and M_l provide scaled replica currents I_h and I_l of the output currents I_H and I_L. Degeneration resistors R_h and R_l have been added to limit the replica currents. The distortion caused by these resistors can be neglected since only the smallest of the two replica currents has to be accurate. The currents I_h and I_l are fed into the block indicated as *quiescent control* which contains a translinear bridge network that is in balance if the currents obey the so-called *harmonic mean* law:

$$\frac{I_h \cdot I_l}{I_h + I_l} = \frac{I_q}{2}$$

(5.2.12)

If the bridge is unbalanced then the current of the two current sources I_1 is adjusted so that the bridge becomes balanced. Consequently, if one of the output transistors is driven very hard, the current in the opposite transistor approaches $I_q/2$ instead of zero.

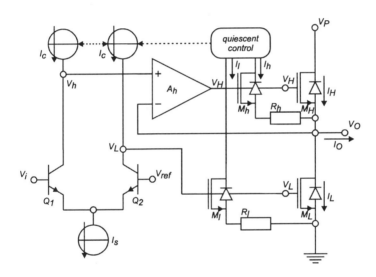

Figure 5.2.7 Minimum selector output stage

Because this mechanism controls the smallest of the currents it is often called the *minimum selector*. The output transistor currents for a geometric mean law and harmonic mean law are shown in Figure 5.2.8.

As can be seen in Figure 5.2.8 the crossover behavior is very smooth and the turn-off of either output transistor is prevented. The amplifier *A* reduces distortion caused by V_{gs} variations of M_H similar to the previous output stage. The voltage gain of this stage is determined by the differential pair Q_1 and Q_2 and higher than unity. The local feedback loop around M_L that was used in the previous stages is still present but is now included in the bias control loop. Consequently, the output resistance of the stage is low.

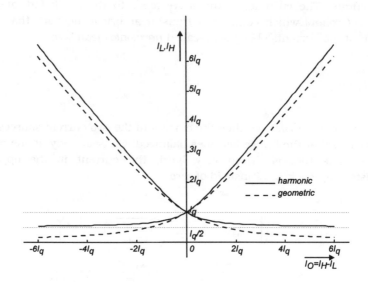

Figure 5.2.8 Output transistor currents as a function of output current

5.2.2 Common Source Stages

Common source output stages are used widely in low voltage CMOS circuits because of their rail-to-rail capability. In a source follower the voltage difference between the output voltage and the supply rails is at least as large as the threshold voltage of the output transistors. In a common source configuration the output is only limited by the on-resistance of the output transistors which can be very low. A distinction can be made between transconductance amplifiers and current amplifiers as shown in Figure 5.2.9.

In the transconductance stage shown in Figure 5.2.9(a) the quiescent current is determined by the replica transistors M_h and M_l. The input voltage V_i is applied to the output transistors M_H and M_L by means of voltage summers Σ_H and Σ_L. To prevent the output transistors from turning off the summers have to be combined with some kind of voltage clipping in a similar way as was done in the floating buffer stage discussed previously.

In the current amplifier stage shown in Figure 5.2.9(b) the input current I_i is fed into a so-called *signal splitter* S that splits the signal into a positive part I_h and negative part I_l, in such a way that $I_i = I_h - I_l$. The signals I_h and I_l are then mirrored and amplified by the mirror ratio N of current mirrors M_h/M_H and M_l/M_L respectively.

If the signal splitter is assumed to be operating correctly, the current transfer I_o/I_i of the current amplifier in Figure 5.2.9(b) is very linear. In contrast, the transconductance stage in Figure 5.2.9(a) has considerable large signal distortion because the nonlinear transfer characteristic of the output transistors is not compensated for.

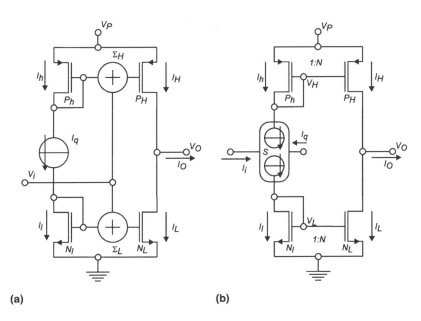

Figure 5.2.9 CMOS common source stages (a) transconductance amplifier (b) current amplifier

Further, the transconductance stage requires some form of clipping to prevent turn-off of the output transistors. In the current amplifier the residual current is controlled by the signal splitter which can be designed in such a way that I_l and I_h obey a *harmonic mean* law such as in the minimum selector stage discussed previously.

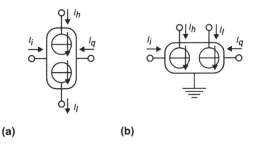

Figure 5.2.10 Signal splitters (a) floating (b) grounded

BCD Audio Amplifiers

In the following it is assumed that a *signal splitter* is available in one of the forms shown in Figure 5.2.10. The splitter shown in Figure 5.2.10(a) is floating and can be connected between the supply and ground. The grounded splitter shown in Figure 5.2.10(b) has both currents flowing in the same direction. The output currents I_h and I_l obey the *harmonic mean law*:

$$\frac{I_h \cdot I_l}{I_h + I_l} = \frac{I_q}{2} \qquad (5.2.13)$$

and:

$$I_h - I_l = I_i \qquad (5.2.14)$$

The distortion of the signal I_h-I_l caused by the splitter is assumed to be negligible compared to the distortion caused by rest of the output stage. The design of a current splitters that meets this profile is discussed in the next section.

Current Amplifier

A straightforward implementation of an all *n-DMOS* current amplifier is shown in Figure 5.2.11.

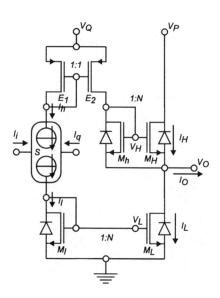

Figure 5.2.11 Current amplifier

Output Stage Topologies

It is composed of two stacked current DMOS mirrors M_h/M_H and M_l/M_L with a large mirror ratio N (~1:1000). The EPMOS mirror E_1/E_2 is needed to mirror the current I_h from the signal splitter S in the required direction and is in first instance assumed to have a mirror ratio 1:1. The operation of this stage is straightforward since there are no local feedback loops present.

However, in this simple form the stage causes unacceptable static and dynamic distortion. The bandwidth of the DMOS mirrors is rather small due to the large mirror ratio N. The large output transistors M_H and M_L have a large gate-source and gate-drain capacitance. At higher frequencies the currents required to charge and discharge these capacitances cause significant dynamic large signal of I_h and I_l. The capacitances can also cause turn-off of the output transistors.

Consider, for example the low-side mirror M_l/M_L with parasitics shown in Figure 5.2.12. During positive excursions of the output voltage V_o the input current of the mirror i_l is very small so that both M_l and M_L are biased at the edge of conduction. Consequently, V_L becomes a high impedance node.

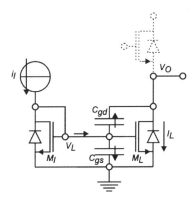

Figure 5.2.12 Parasitic capacitances in DMOS mirror

If the output voltage V_o drops the voltage V_L is also forced down by the capacitive coupling through C_{gd} and C_{gs}.. This can turn off M_l and M_L which may cause dynamic crossover distortion. This situation is similar for the high-side mirror M_h/M_H. The influence of the parasitic capacitances can be reduced by including a voltage buffer in the DMOS mirrors. Two simple alternatives are shown in Figure 5.2.13.

The configuration shown in Figure 5.2.13(a) has the advantage that the gate-source voltages of M_l and M_L are exactly equal but has a local feedback loop around M_l. Especially in the floating high-side mirror the stability of this loop may be a problem because of the large variations in the transconductance of M_l.

The configuration shown in Figure 5.2.13(b) does not have a local feedback loop but has the disadvantage that the offset voltage of the buffer G_l causes a voltage error between the gate-source voltages of M_l and M_L. This offset should preferably be smaller than the threshold voltage mismatch of the DMOS transistors.

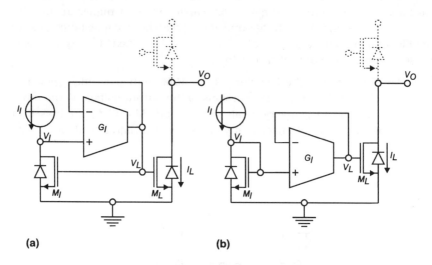

Figure 5.2.13 DMOS mirrors with voltage buffer

More problems with the stage shown in Figure 5.2.11 arise around the EPMOS mirror E_1/E_2. First, the high-side DMOS current mirror M_h/M_H has its common source connected to the output V_o. Consequently, the input node of the mirror V_H follows the voltage swing of the output. Due to the finite output impedance of the EPMOS mirror E_1/E_2 this may result in considerable static large signal distortion of I_h. Second, in order to have rail-to-rail output capability, the EPMOS mirror E_1/E_2 has its common source connected to V_Q. This voltage is generated by a bootstrap or a chargepump circuit neither of which behaves as a steady supply rail. Further, if a chargepump is used, the maximum current that can be mirrored is limited by the current capability of the chargepump.

These problems can be solved by using the alternative buffering for the high-side DMOS mirror shown in Figure 5.2.14. The transconductance amplifier G_h forces the voltage V_h equal to a voltage V_{ref} lower than the supply voltage V_P. Consequently, the output voltage swing is no longer present at this node. Because V_h is lower than the supply voltage, the EPMOS mirror can be connected to the supply V_P instead of V_Q. However, the transconductor G_h still needs to be supplied by V_Q. Again the local feedback loop around M_h may cause instability.

An output stage using the configurations shown in Figure 5.2.13(a) and Figure 5.2.14 for the low-side and high-side DMOS mirrors respectively was studied in [16]. It appeared to be very difficult to guarantee the stability of this output stage due to the local feedback loop around M_H.

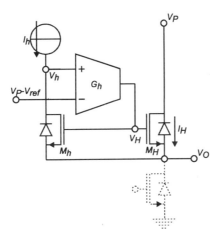

Figure 5.2.14 Alternative buffer for high-side mirror

Improved Current Amplifier

A simpler method to solve the problems caused by the *p*-mirror is shown in Figure 5.2.15. In this stage, the *p*-mirror has been removed completely while the high-side current mirror is replaced by the combination of M_h and M_H and two transconductors G_{hx} and G_{hy}. The transconductors G_{hx} and G_{hy} copy the voltage difference between the gate and source of M_h between the gate and source of M_H. The gate of DMOS M_h can be connected to a constant reference voltage V_{ref} and does not follow the output voltage V_o anymore. The combination M_h, M_H, G_{hx} and G_{hy} operates as a current amplifier with floating input and output terminals. The same method is used to buffer the low-side mirror with transconductors G_{lx} and G_{ly}.

The control of the currents through the output transistor M_H and M_L is completely feedforward and no local feedback loops including one of the DMOS output transistors is used. Consequently, the stability of this stage can be expected to be very good.

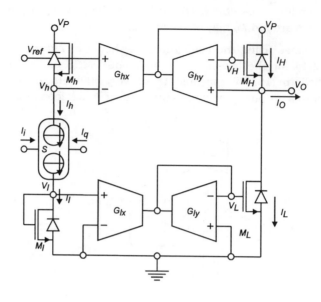

Figure 5.2.15 Improved current amplifier

The stage can also easily be adapted for a grounded signal splitter as shown in Figure 5.2.16. As is explained later, a grounded signal splitter is preferred over a floating version.

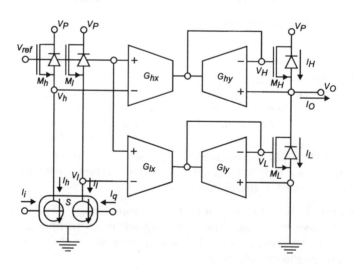

Figure 5.2.16 Improved current amplifier with grounded splitter

Minimum Selector Transconductance Amplifier

The minimum selector output stage presented earlier as a voltage buffer can also be modified into a transconductance amplifier. This results in the circuit shown in Figure 5.2.17.

The difference with the voltage buffer version lies in the different way of driving the high-side transistor M_H. In the voltage buffer, the gate of M_H is driven by amplifier A in such a way that the output voltage V_o becomes equal to voltage V_h. In the transconductance amplifier the voltage V_h is copied between the gate and source of M_H independent of V_o. Actually, this is an implementation of a transconductance amplifier as was shown in Figure 5.2.9(a). Similar to the voltage buffer version, the currents I_H and I_L obey a harmonic mean law so there is no need for clamping of the gate-source voltages of the output transistors.

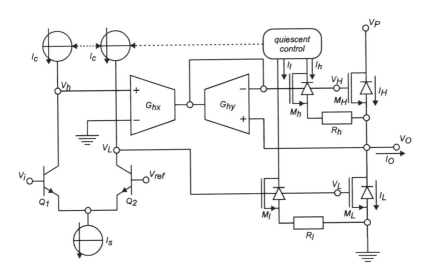

Figure 5.2.17 Minimum selector as a transconductance amplifier

5.2.3 Comparison of Output Stages

All output stages described in the previous section have in common that one or two scaled replicas of the output transistors are used to control the quiescent current. However, the exact role of the replicas is different in the various stages. This has consequences for the ratio, and thus the device and temperature matching, between the replicas and the output transistors.

In the *quasi-complementary voltage buffers* the scaled replicas conduct a constant current which is used as a reference for the quiescent current in the output transistors. In these stages the ratio between replica and output transistors can be chosen relatively small (e.g. *1:20*) resulting in good matching and thus accurate control of the quiescent current.

In the *minimum selector stages*, the replicas are used to generate replica currents of the currents through the output transistors which are then used in a feedback loop that controls the residual current in the output transistor that is not conducting the signal current. The accuracy of this control loop is limited by the accuracy of the replica currents. However, the maximum current in the output transistors can be as high as *10A*. In order to limit the replica currents to acceptable levels the ratio between replica and output transistors has to be chosen larger than in the quasi-complementary voltage buffer stages (e.g. *1:100*).

In the *current amplifier stages* a multiple of the current through the replicas flows through the output transistors. The current amplification factor is determined by the ratio between replica and output transistors and is typically very large (e.g. *1:1000*).

The *quasi-complementary voltage buffers* have in common that they all require some form of voltage clamping to prevent turn-off of the output transistors. The time that is needed to recover from clamping may result in severe dynamic crossover distortion or even instability. Further, these stages have in common that a local feedback loop is used including the low-side output transistor M_L. The large variations in the output current result in large variations in the transconductance of the output transistors and thus in large variations in the loopgain of this local feedback loop. In combination with the rather large input capacitance of the output transistors it is difficult to keep this local loop stable.

The *minimum selector stages* are expected to show smooth crossover due to the harmonic mean relation that is forced on the output transistors. However, the local feedback loop may also cause problems with the stability of these stages.

In the *improved current amplifier stage* no local feedback is used around the output transistors. The quiescent and residual currents are controlled by the signal splitter in a straightforward manner. Further, large signal distortion is expected to be low due to the compensating effect of the current mirrors. A disadvantage of this stage is that it may be difficult to achieve proper matching between the replica and output transistors due to the large ratio that is required for current amplification.

Summarizing, the *quasi-complementary voltage buffers* are expected to have accurate quiescent current control but poor crossover behavior and stability. The *minimum selector stages* are expected to have smooth crossover behavior, reasonably accurate quiescent current control but poor stability. The *improved current amplifier stages* are expected to have good stability, low large signal and crossover distortion but quiescent current control may not be very accurate due to the poor matching of the output and replica transistors.

5.3 A BCD Amplifier Design

In this section the design and implementation of an audio amplifier in a BCD technology is described. In the previous section a comparison has been made between various different output stage topologies.

Based on this comparison a combination of a *signal splitter* with an *improved current amplifier* is selected for further investigation. The main reason for selecting this stage is the superior stability that can be expected because no local feedback loops around the output transistors are used.

5.3.1 Amplifier Topology

Since the output stage is a current amplifier, the complete amplifier topology becomes a transconductance amplifier as was described in Section 5.1.2. For clarity this topology is shown again in Figure 5.3.1.

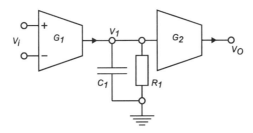

Figure 5.3.1 Transconductance amplifier topology

Before an implementation of the different stages can be designed, some calculations have to be made for a rough estimate of the specifications for each stage. For each stage it has to be determined what the required transconductance or current gain is and how much bias current is available to realize this. Some specifications for the complete amplifier are described in Section 1.1.3.

Closed Loop Voltage Gain and Output Resistance

The closed loop voltage gain of an audio power amplifier usually lies between *26dB* and *30dB*. In Section 5.1.2 it is shown that the closed loop voltage gain is equal to $1/\beta$. The closed loop output resistance R'_o is given by:

$$R'_o \approx \frac{1}{G_m \cdot \beta}$$

(5.3.15)

The output resistance typically has a value between $10m\Omega$ and $100m\Omega$. Consequently the total transconductance G_m lies between $300A/V$ and $3000 A/V$.

The total available bias current at zero output current is set at $25mA$. The quiescent current I_q in the output transistors is chosen to be $10mA$ which leaves $15mA$ for the output transistor drivers, the signal splitter and the input stage. The supply voltage V_P of the amplifier is $60V$ resulting in a quiescent dissipation P_q of $1.5W$ and maximum output power $P_{o,max}$ of about $110W/4\Omega$.

Output Transistor Dimensions

The characteristics of the output transistors determine to a large extent the specifications for the other stages. In order to minimize the area consumption and input capacitance, the size of the output transistors should be as small as possible. However, the minimum size is limited by requirements for the on-resistance and the thermal resistance to the heatsink.

The *on-resistance* R_{on} of the output transistors limits the output voltage window and thus the maximum output power of the amplifier.

The *thermal resistance* R_{th} to the heatsink of the output transistors relates the transistor temperature to the power dissipation in the transistor. A higher thermal resistance results in a higher temperature. This also limits the maximum output power of the amplifier.

Both *on-resistance* and *thermal resistance* are roughly inversely proportional to the transistor area. For this design the area of the output transistors is approximately $4mm^2$ each. The *on-resistance* R_{on} for these transistors is about $200m\Omega$. With a 4Ω load resistor the maximum output voltage swing is reduced by about $1.5V$ on both sides resulting in a reduction of the maximum output power to $100W/4\Omega$. The thermal resistance R_{th} of the output transistors to the heatspreader in the power package is approximately $0.3K/W$. With a 4Ω load resistor the maximum instantaneous dissipation in the output transistors is $56W$ resulting in a temperature rise of $17K$ with respect to the heatspreader temperature which is a safe value.

Output Transistor Drivers

Each output transistor is driven by a combination of two transconductance amplifiers that copy the gate-source voltage of the replica transistors to the gate-source voltage of the output transistors. These drivers have to be able to supply the current needed to charge and discharge the gate capacitance of the output transistors. The magnitude of these currents depends on the input capacitance and transconductance of the output transistors, the load current and the slope of the output voltage. It is assumed that both drivers have class A operation which means that the output current of the drivers is always smaller than their bias current. Further, it is assumed that the high-side driver uses the output of a chargepump as a supply voltage V_Q. As is explained in Chapter 3, a chargepump consumes at least twice the current it can supply. Consequently, the maximum output current of each of the output transistor drivers is smaller than one third of the bias current consumption of both output drivers and chargepump.

Peak currents occur during crossover if the transconductance of the output transistors is lowest. In Section 1.1.3 it is calculated that the lowest value for the slew-rate SR of the amplifier is about $4V/\mu s$. In order to estimate the magnitude of the peak currents a small signal equivalent circuit of the output stage in the crossover region can be used as shown Figure 5.3.2.

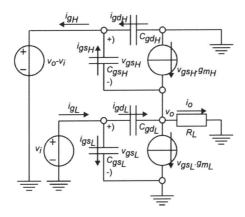

Figure 5.3.2 Small signal equivalent in crossover region

Note that during crossover, both output transistors contribute to the output signal and are driven with opposite signal voltages v_i. With a resistive load R_L and a constant slope $dV_o/dt=SR$ at the output node the capacitive gate current $i_g=i_{gs}+i_{gd}$ for the output transistors can be approximated by:

$$i_g(t) = \frac{C_{gs} + C_{gd}}{2 \cdot g_m \cdot R_L} \cdot SR \cdot \left(1 - e^{-\frac{t}{\tau}}\right) + C_{gd} \cdot SR \text{ with } \tau = \frac{2 \cdot g_m}{C_{gs} - C_{gd}}$$

(5.3.16)

for $t >> \tau$ this expression converges to the maximum value:

$$i_{g,max} = \left(\frac{C_{gs} + C_{gd}}{2 \cdot g_m \cdot R_L} + C_{gd}\right) \cdot SR$$

(5.3.17)

In the crossover region the current through both output transistors is equal to the quiescent current I_Q which equals *10mA*.

At a draincurrent I_D of *10mA* the values of the transconductance g_m, gate-source capacitance C_{gs} and the gate-drain capacitance C_{gd} are approximately *100mA/V*, *200pF* and *20pF* respectively. With a *4Ω* load resistor this leads to a maximum gate current $I_{g,max}$ of about *1mA*. In practice the maximum current is smaller since the transconductance g_m increases rapidly. For drain currents smaller than about *100mA* the output transistors are in weak inversion and the transconductance g_m is proportional to the drain current I_D.

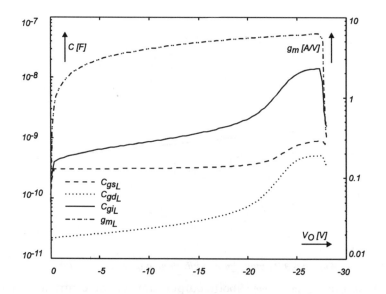

Figure 5.3.3 *Output voltage dependence Transconductance g_m and capacitances C_{gsL}, C_{gdL} and C_{giL}*

High gate current may also occur during large excursions of the output signal V_o. Due to the accumulation effect described in Chapter 2 the gate-source and gate-drain capacitances C_{gs} and C_{gd} increase substantially if the gate voltage approaches the drain voltage. A (logarithmic) plot of C_{gsL} and C_{gdL} of the low-side transistor M_L as a function of V_o with a 4Ω load resistor is shown in Figure 5.3.3. As can be seen in Figure 5.3.3 the gate-source capacitance C_{gsL} ranges from about *300pF* in the crossover region up to almost *1nF* for large output voltage excursions. However, more significant is the change in C_{gdL} which ranges from about *20pF* in the crossover region to about *500pF* for large output voltage excursions.

Although the value of C_{gdL} is smaller than C_{gsL} it appears much larger due to the Miller-effect. Consider the small signal equivalent of the output stage during negative output voltage excursions shown in Figure 5.3.4. During large negative output voltage excursions the transconductance g_{mL} of the low-side output transistor M_L is much larger than in the crossover region. In this case the voltage transfer can be approximated as:

$$v_o = -g_{mL} \cdot R_L \cdot v_i$$

(5.3.18)

and the input capacitance C_{giL} is then given by:

$$C_{giL} = C_{gsL} + (1 + g_{mL} \cdot R_L) \cdot C_{gdL}$$

(5.3.19)

The maximum value of g_{mL} is about *7A/V* so with a 4Ω load resistor R_L the value of C_{gdL} is multiplied by a maximum value of *29*.

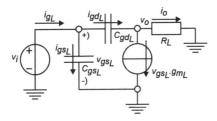

Figure 5.3.4 Small signal equivalent during negative output voltage excursions

As can be seen in Figure 5.3.3, for large output voltage excursions the input capacitance C_{giL} is completely dominated by the Miller-effect. The maximum value of C_{giL} is almost *15nF*.

The capacitive gate current i_{gL} can be approximated with:

$$i_{gL} = \frac{C_{giL}}{g_{mL} \cdot R_L} \cdot \frac{dv_O}{dt}$$

(5.3.20)

The dV_o/dt of a 20kHz full swing sinewave at an output voltage of V_o= -25V is about ±2V/µs while the input capacitance C_{giL} at V_o= -25V is 14nF. This results in a capacitive input current i_{gL} of about 1mA.

The same holds for the high-side transistor M_H during positive voltage excursions. Although, the drain of the high-side transistor M_H is connected to the constant supply voltage V_p, the Miller-effect is still present because the source of M_H follows the output signal V_o. Consequently, V_{gs} and V_{ds} of M_H during positive output excursions are almost identical to V_{gs} and V_{ds} of M_L during negative excursions.

Based on the previous a quiescent current of 10mA is set for the transistor drivers which means that the maximum output current for each driver will be in the region of 2.5mA to 3.0mA. This leaves 5mA for the signal splitter and input stage.

Replica Transistors

The current gain F of the output stage is set by the ratio of the gate widths of the output transistors and the replica transistors. As is explained in Chapter 2, VDMOS transistors are composed of a number of small cells connected in parallel. The selected output transistors have 20 fingers with 206 cells each while the smallest VDMOS transistor has 5 cells. The resulting current gain F is then equal to 824 which is too low. An output current of 10A would require an input current of 12mA which has to be supplied by the signal splitter and input stage. The available bias current for these stages is only 5mA which means that class A operation would not be possible.

A higher current gain can be achieved if LDMOS instead of VDMOS transistors are used as replicas. The channel of LDMOS transistors is made in the same way as that of VDMOS transistors so their threshold voltage and channel length are equal. The gate width of one VDMOS cell is approximately 50µm. Consequently, if an LDMOS transistor with a gate width of 25µm is used as replica transistor, the current gain, F becomes 8240.

A possible drawback of this solution is that the matching between the VDMOS and LDMOS transistors may be relatively poor.

Signal Splitter and Input Stage

The transconductance output stage G_2 in Figure 5.3.1 is composed of a transconductor G_3, a signal splitter S and two current amplifiers F_H and F_L, one for the high-side and one for the low-side as shown in Figure 5.3.5. The transconductance G_2 of the complete output stage is then given by:

$$G_2 = F_{H,L} \cdot G_3$$

(5.3.21)

A BCD Amplifier Design

The input stage has a transconductance G_1 and is loaded with a resistor R_1. Consequently, the transconductance G_m of the complete amplifier is:

$$G_m = G_1 \cdot R_1 \cdot F_{H,L} \cdot G_3$$

(5.3.22)

The resistor R_1 is connected in parallel with a capacitor C_1 in order to create a dominant pole. Usually, the dominant pole frequency in audio amplifiers is in the range of *1kHz* to *10kHz*. A dominant pole frequency of about *5kHz* is realized by choosing $R_1=200k\Omega$ and $C_1=150pF$. This combination results in reasonable area consumption for this *RC* combination. Further, the value of G_1 and G_3 are chosen *400µA/V* and *1mA/V* respectively resulting in an overall G_m of *824* which is in the required range. These values appear to be rather arbitrary but have been optimized during the design process to yield a good compromise between loop gain, dominant pole frequency, open loop distortion and stability requirements.

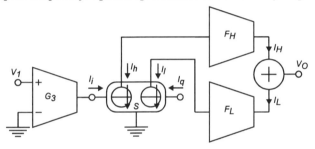

Figure 5.3.5 Output stage topology

A detailed description of the different amplifier stages is presented in the following sections.

5.3.2 Signal Splitter

The signal splitter *S* has a voltage V_1 as input and two currents I_h and I_l as outputs. The currents are amplified separately by output current amplifiers and then summed at the output node V_o of the amplifier. In order to achieve low crossover distortion at the output the crossover distortion caused by the signal splitter should be as low as possible. Crossover distortion of the signal splitter can be determined by considering the difference between the output currents I_h-I_l. Although large signal distortion is reduced by the loop gain of the complete amplifier it desirable to minimize the large signal distortion caused by the signal splitter as well. A very smooth crossover behavior can be achieved if the output currents obey the *harmonic mean* law:

BCD Audio Amplifiers

$$\frac{I_h \cdot I_l}{I_h + I_l} = \frac{I_q}{2}$$

(5.3.23)

which has the advantage that the smallest of the two currents has a lower bound at $I_q/2$ and thus prevents the output transistors from turning off. This harmonic mean law can be realized by using a translinear loop. The quiescent current I_Q in the output transistors is set at *10mA* while the current gain $F_{H,L}$ of the current amplifiers is *8240*. Consequently, the quiescent current I_q of the signal splitter becomes *1.2μA*.

A distinction can be made between *passive* and *active* signal splitters.

Passive Signal Splitters

A simple implementation of a passive floating signal splitter is shown in Figure 5.3.6(a). This is basically the same circuit as the complementary voltage buffer shown earlier in Figure 5.2.2(a) but the output is now used as input.

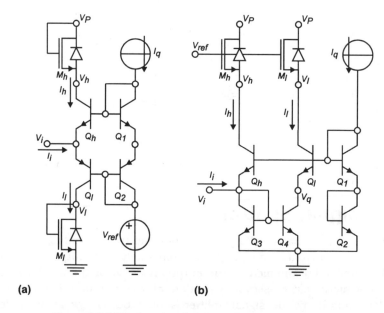

Figure 5.3.6 Passive signal splitters (a) floating (b) grounded

Due to the translinear loop formed by transistors Q_1, Q_2 Q_l and Q_h the output currents I_h and I_l obey the *geometric mean* law:

$$I_h \cdot I_l = I_q^2$$

(5.3.24)

A BCD Amplifier Design

However, the performance of this splitter is deteriorated by the PNP transistors. In the available BCD technology only *lateral PNP* transistors are possible. These transistors have a low f_T (~5MHz) and the current gain β_F ranges from 10 to 100 depending on the collector current. This leads to considerable large signal distortion in the low-side signal I_l.

The NPN transistors perform much better due to their vertical structure. The *vertical NPN* transistors have an f_T around 300MHz and an almost constant current gain β_F of 60. A grounded signal splitter using only NPN transistors is shown in Figure 5.3.6(b). A translinear loop is formed by transistors Q_1, Q_2, Q_h and Q_3. that causes the collector currents of these transistor to obey the *geometric mean* law. The current through Q_3 is mirrored by Q_4 and is cascoded by Q_l. However, this signal splitter also has some drawbacks.

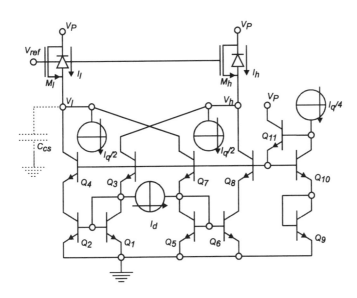

Figure 5.3.7 Improved grounded signal splitter

Due to the base currents of Q_3 and Q_4 the mirror ratio is less than unity. As a result, the low-side half of the signal is attenuated slightly which results in even order harmonic distortion. This can be compensated for by using two signal splitters in a differential form. In this case the input current also has to be available differentially. Further, when the base current of Q_h or Q_l exceeds the value of I_q this causes one or more transistor in the translinear loop to turn off which results in dynamic crossover distortion. This can be solved by buffering the base currents of Q_h and Q_l. Finally, the output currents obey a *geometric mean law* whereas

BCD Audio Amplifiers

a *harmonic mean law* is preferred. The *geometric mean law* relation can easily be transformed into a *harmonic mean law* by reducing the quiescent current to $I_q/2$ and adding a constant current $I_q/2$ to both output currents. An improved all-NPN signal splitter is shown in Figure 5.3.7.

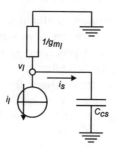

Figure 5.3.8 Small signal equivalent of Q_i and M_h.

Vertical NPN transistors have a large collector-substrate capacitance C_{cs} compared to other parasitics due to their heavily doped buried collector diffusion. For a minimum size transistor the value of C_{cs} is about *400fF*. Further, the voltages on the collectors of Q_1, through Q_8 in Figure 5.3.7 are highly nonlinear. The capacitive currents through the collector-substrate capacitances are largest during crossover where the slope of the collector voltages is largest and the current smallest. For example, consider transistors Q_4, Q_7 and M_l. A small signal equivalent circuit is shown in Figure 5.3.8 where C_{cs} is the combined collector-substrate capacitance of Q_4 and Q_7, g_m is the transconductance of M_l and i_l is the (differential) input current i_d. If the differential input current i_d is a sinewave with frequency f and amplitude I_d, the di_d/dt during crossover is equal to $2\pi f I_d$. The capacitive current i_s through C_{cs} is then given by:

$$i_s(t) = \frac{C_{cs}}{g_{ml}} \cdot 2\pi \cdot f \cdot I_d \cdot \left(e^{-\frac{t}{\tau}} - 1\right) \text{ with } \tau = \frac{C_{cs}}{g_{ml}}$$

(5.3.25)

In the crossover region the current through M_l is equal to the quiescent value I_q which is set at *1.2µA*. The value of g_{ml} is then approximately *12µA/V*. For a *1kHz* sinewave with a *500µA* amplitude the value of i_s converges to about *200nA* which is *18%* of the quiescent current. The currents through the collector-substrate capacitances of the other vertical NPN transistors have comparable values. These currents result in considerable dynamic crossover distortion.

Active Signal Splitter

With the passive signal splitters discussed in the previous it is only possible to realize a geometric mean law directly which can be transformed into a harmonic mean law by adding a constant current to both I_h and I_l. A direct realization of the harmonic mean law requires a more complex technique. Consider, the situation shown in Figure 5.3.9.

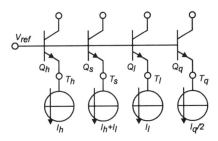

Figure 5.3.9 Indirect translinear loop

If the currents I_h and I_l obey the harmonic mean law then the emitter voltages of transistors Q_l, Q_s, Q_h and Q_q have a translinear relation:

$$V(T_h) + V(T_l) - V(T_s) - V(T_q) =$$
$$-\frac{kT}{q} \cdot \ln(I_h) - \frac{kT}{q} \cdot \ln(I_l) + \frac{kT}{q} \cdot \ln(I_h + I_l) + \frac{kT}{q} \cdot \ln(I_q/2) =$$
$$\frac{kT}{q} \cdot \ln\left(\frac{(I_h + I_l) \cdot I_q/2}{I_h \cdot I_l}\right) = 0$$

(5.3.26)

In other words, if the emitter voltages do not obey this relation, the difference voltages between the emitters can be used in a feedback loop that controls I_h and I_l. A principle schematic of this feedback loop is shown Figure 5.3.10. In Figure 5.3.10 the floating current source I_d represents the (differential) input signal. Both input currents I_d and $-I_d$ are summed with a common current I_c resulting in the high-side and low-side currents I_h and I_l:

$$I_h = I_c + I_d$$
$$I_l = I_c - I_d$$

(5.3.27)

The currents I_h and I_l are split into two equal halves by the transistors M_h and M_l and then led to the emitters of the bipolar transistors Q_h, Q_s and Q_l, while transistor Q_q conducts a constant current $I_q/4$. The four-input

BCD Audio Amplifiers

common mode transconductance amplifier g_c controls the common current I_c:

$$I_c = g_c \cdot \left(V(T_q) + V(T_s) - V(T_h) - V(T_l)\right)$$

(5.3.28)

Consequently, if the transconductance g_c is very large then the relation between the currents I_h and I_l approximates the harmonic mean law. Note that the transistors M_h and M_l not only operate as current splitters but at the same time they are the replica transistors. Therefore, the collectors of the bipolar transistors can be connected to the supply voltage V_P which eliminates the influence of capacitive collector-substrate currents.

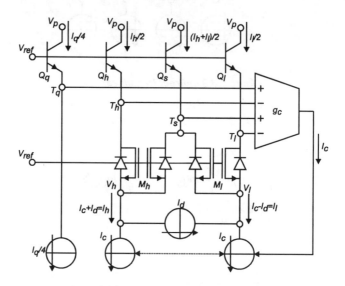

Figure 5.3.10 Active feedback loop

The drain-substrate of the replica transistors is much smaller because they do not have a buried layer. The gates of M_h and M_l can be connected to the same voltage V_{ref} as the bases of Q_q, Q_h, Q_s and Q_l. Since the threshold voltage V_T (~2.5V) of the DMOS transistors is much larger than one V_{be} (~0.6) of a bipolar transistor, transistors M_h and M_l are always saturated. The influence of base currents is also completely eliminated. The loop has perfect symmetry which suppresses even order harmonic distortion.

A BCD Amplifier Design

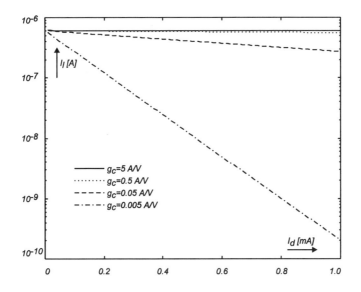

Figure 5.3.11 Residual current for different values of g_c

As mentioned before, the currents I_l and I_h obey the harmonic mean law if the transconductance g_c is very large. However, a large value of g_c results in a large loop gain and this may endanger the stability of the loop. On the other hand, if the transconductance g_c is small an error voltage I_c/g_c appears at the input which disturbs the translinear relation (5.3.26). The relation between I_d and I_c can be expressed as:

$$I_c = g_c \cdot (V(T_q) + V(T_s) - V(T_h) - V(T_l)) \Rightarrow I_c = g_c \cdot \frac{kT}{q} \cdot \ln\left(\frac{I_c \cdot I_q}{I_c^2 - I_d^2}\right) \Rightarrow$$

$$I_d^2 = I_c^2 - I_c \cdot I_q \cdot e^{-\frac{I_c}{g_c} \cdot \frac{q}{kT}}$$

(5.3.29)

whereas for the ideal harmonic mean the relation between I_d and I_c can be calculated as:

$$\left.\begin{array}{l} I_h \cdot I_l = (I_h + I_l) \cdot I_q/2 \\ I_h = I_c + I_d \\ I_l = I_c - I_d \\ I_d^2 = I_c^2 - I_c \cdot I_q \end{array}\right\} \Rightarrow$$

(5.3.30)

BCD Audio Amplifiers

The difference between (5.3.29) and (5.3.30) is the exponential term. For large values of g_c this term approaches unity but for small values the influence increases. The effect of the exponential term on the residual current is shown in Figure 5.3.11 where the low-side current I_l is shown as a function of I_d for different values of g_c. For negative I_d the same applies to the high-side current I_h.

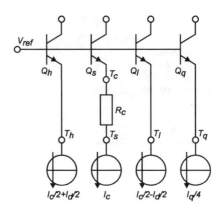

Figure 5.3.12 Series resistor R_c

For small g_c the residual current becomes much smaller than the desired value $I_q/2$ which is unacceptable, so a large g_c seems highly desirable. However, the influence of the exponential term in (5.3.29) can also be reduced by including a small series resistor R_c in series with the emitter of transistor Q_s as shown in Figure 5.3.12. Due to this simple modification the relation between I_d and I_c now becomes:

$$I_c = g_c \cdot (V(T_q) + V(T_s) - V(T_h) - V(T_l)) \Rightarrow$$

$$I_c = g_c \cdot \left(\frac{kT}{q} \cdot \ln\left(\frac{I_c \cdot I_q}{I_c^2 - I_d^2}\right) + R_c \cdot I_c \right) \Rightarrow$$

$$I_d^2 = I_c^2 - I_c \cdot I_q \cdot e^{-\frac{I_c \cdot (1 - g_c \cdot R_c)}{g_c} \cdot \frac{q}{kT}}$$

(5.3.31)

So, if the value of R_c equals $1/g_c$ the exponential term vanishes and the relation between I_d and I_c corresponds to the ideal harmonic mean law, independent of the value of g_c. A small signal expression for i_c/i_d is given by:

A BCD Amplifier Design

$$i_c = g_c \cdot \left((r_s + R_c) \cdot i_c - r_h \cdot (i_c + i_d)/2 - r_l \cdot (i_c - i_d)/2\right) \Rightarrow$$

$$\frac{i_c}{i_d} = \frac{(r_l - r_h)/2}{1 + g_c \cdot ((r_l + r_h)/2 - r_s) - g_c \cdot R_c}$$

(5.3.32)

where r_h, r_l and r_s are the inverse transconductances of transistors Q_h, Q_l and Q_s respectively:

$$r_h = 1/g_{mh} = \frac{kT}{q \cdot (I_c + I_d)/2} \qquad r_l = 1/g_{ml} = \frac{kT}{q \cdot (I_c - I_d)/2}$$

$$r_s = 1/g_{ms} = \frac{kT}{qI_c}$$

(5.3.33)

The loopgain A_β of the feedback loop is given by:

$$A_\beta = g_c \cdot ((r_l + r_h)/2 - r_s) - g_c \cdot R_c$$

(5.3.34)

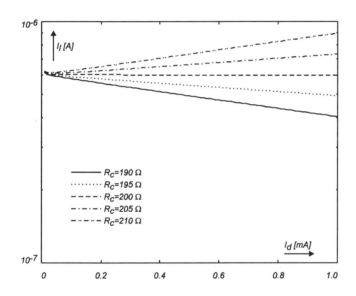

Figure 5.3.13 Residual current for different values of R_c

The value of $(r_l + r_h)/2 - r_s$ is equal to kT/qI_q for $I_d = 0$ and approaches $2kT/qI_q$ for $I_d <> 0$. For $I_q = 1.2\mu A$ and $kT/q = 25mV$ these values are $20k\Omega$ and $40k\Omega$ respectively. If a moderate loopgain A_β, for example between 40dB and 48dB, is chosen the corresponding value for g_c is about 5mA/V and the ideal value for R_c becomes 200Ω. The influence of R_c on the loopgain A_β is negligible. The effect of different values for R_c and $g_c=5mA/V$ on the residual current is shown in Figure 5.3.13. As can be seen in Figure 5.3.13 the residual current tends to increase with I_d if the product g_cR_c is greater than unity. Consequently, it is better to choose the value of R_c slightly smaller than the ideal value in order to leave some room for mismatch.

Implementation

The differential input current source I_d is implemented as a differential mode transconductance amplifier shown in Figure 5.3.14. It is composed of a linearized degenerated differential EPMOS pair E_{d1}, E_{d2}. The feedback loops formed by DMOS transistors M_{d1} and M_{d2} force E_{d1} and E_{d2} to conduct a constant current I_2. Therefore, the input voltage V_i is copied almost distortionless to the conversion resistor R_d resulting in a very linear current I_d/m. This current I_d/m is superimposed on the bias draincurrents of M_{d1} and M_{d2}. These draincurrents are mirrored and multiplied by m by DMOS transistors M_{d3} and M_{d4}. The transfer function of this circuit is given by:

$$I_d = \frac{m}{R_d} \cdot V_d = g_d \cdot V_d$$

(5.3.35)

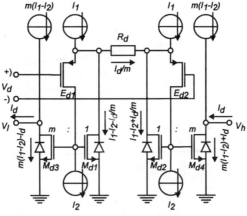

Figure 5.3.14 Differential mode transconductance amplifier

In the final realization the following values have been used: $R_d=20k\Omega$, $m=10$, $I_2=60\mu A$ and $I_1=120\mu A$. The common mode transconductance amplifier shown in Figure 5.3.15 is implemented using the same linearization technique as the differential mode transconductance amplifier. The voltage differences $V(T_s)-V(T_q)$, $V(T_h)-V(T_q)$ and $V(T_l)-V(T_q)$ are converted into the currents I_s, I_h and I_l by resistors R_s, R_h and R_l respectively. These signal currents are superimposed on the draincurrent of M_{c6} and mirrored and

A BCD Amplifier Design

multiplied by n by M_{c7} and M_{c8}. If the resistors R_s, R_h and R_l have the same value n/g_c then the transfer of this circuit is given by:

$$I_c = -\frac{n}{R_h} \cdot (V(T_h) - V(T_q)) - \frac{n}{R_l} \cdot (V(T_l) - V(T_q)) + \frac{n}{R_s} \cdot (V(T_s) - V(T_q))$$

$$= g_c \cdot (V(T_s) + V(T_q) - V(T_h) - V(T_s))$$

(5.3.36)

In the final realization the following values have been used: $R_{s,h,l}=1k\Omega$, $n=6$, $I_1=720\mu A$, $I_2=60\mu A$, $I_3=120\mu A$ and $I_4=240\mu A$. The resulting g_c for this stage is then $6mA/V$. The emitter-resistor R_c has been set at 160Ω which is a little lower than the optimal value but leaves some room for mismatch between g_c and R_c. Distortion in the current I_c does not affect the output signal of the complete signal splitter since it is common to both I_h and I_l and consequently cancels out in $I_o=I_h-I_l$.

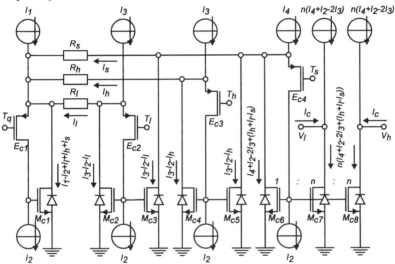

Figure 5.3.15 Common mode transconductance amplifier

Influence of Offset

Both the differential mode and the common mode transconductance amplifiers may introduce offset currents at their outputs as a result of mismatch between transistors in both circuits. However, the common part of these offset currents can be translated to an offset voltage at the input of the common mode transconductance amplifier while the differential part of the offset currents can be translated to an offset voltage at the input of the differential mode transconductance amplifier.

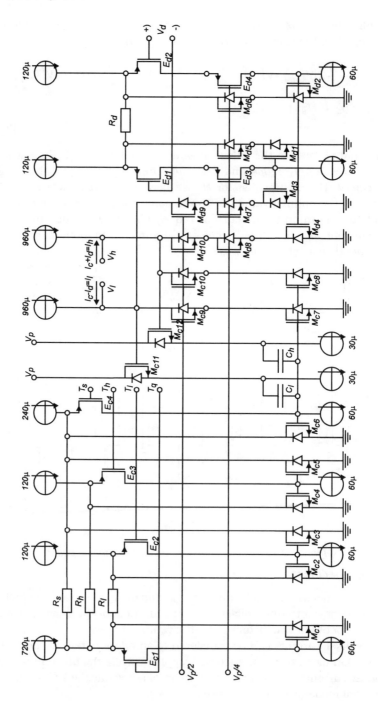

Figure 5.3.16 Differential and common mode transconductance amplifiers

A BCD Amplifier Design

A offset voltage V_{offset} at the input of the common mode transconductance amplifier results in a deviation of the quiescent current value I_q of I_h and I_l:

$$I_q = I_{q0} \cdot e^{\frac{q}{kT} \cdot V_{offset}}$$

(5.3.37)

At room temperature $5mV$ offset results in a 20% variation in quiescent current.

Complete Signal Splitter

A more detailed schematic of the differential and common mode transconductance amplifiers is shown in Figure 5.3.16. In the differential mode transconductance amplifier, cascodes have been added to the EPMOS and DMOS transistors in order to increase the linearity of this stage. Cascodes have also been added to M_{c7} and M_{c8} in the common mode transconductance amplifier in order to increase the output impedance of the stage.

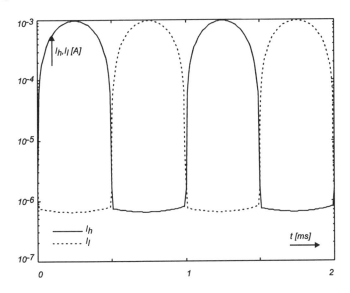

Figure 5.3.17 High-side and low-side currents I_h and I_l

The capacitors C_h and C_l provide frequency compensation of the common mode loop. The source followers M_{c11} and M_{c12} buffer prevent capacitive currents through C_h and C_l to distort the output signal. The cascodes M_{c9}, M_{c10}, M_{d9} and M_{d10} are minimum size DMOS transistors which have been added to minimize the parasitic drain-substrate capacitance at the nodes

BCD Audio Amplifiers

V_h and V_l. The current consumption of the complete circuit is about *3.5mA*. The overall transconductance G_3 of the splitter is:

$$G_3 = \frac{I_o}{V_d} = \frac{I_h - I_l}{V_d} = 2 \cdot \frac{I_d}{V_d} = 2 \cdot g_d = 1mA/V$$

(5.3.38)

The loopgain A_β in the common mode loop ranges from *40dB* to *48dB*.

Simulation Results

The currents I_l and I_h for a sinewave input voltage with an amplitude of *1V* and frequency of *1kHz* are shown in Figure 5.3.17. As can be seen in Figure 5.3.17 the residual value of both currents are limited to about *500nA*.

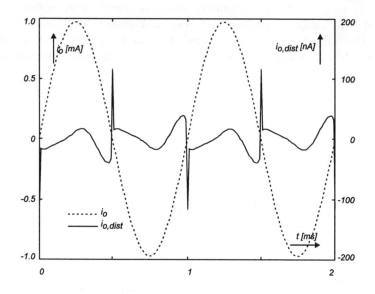

Figure 5.3.18 Distortion in output current $I_o = I_h - I_l$

In Figure 5.3.18 the distortion of the output current $I_o = I_h - I_l$ for the same input signal is shown for the active signal splitter. This distortion is determined by subtracting the first harmonic from the output signal.

For comparison the distortion of the passive signal of Figure 5.3.7 is shown in Figure 5.3.19 together with the distortion of the active signal splitter for the same *1kHz/1V* sinewave signal.

As can be seen both large signal and crossover distortion have been reduced substantially in the active signal splitter.

A BCD Amplifier Design

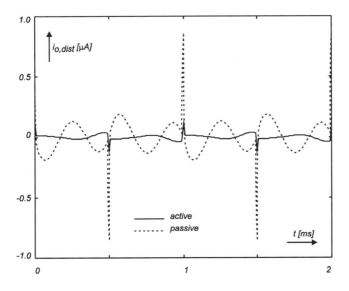

Figure 5.3.19 Comparison of passive and active signal splitter

The total harmonic distortion (THD) as a function of output current amplitude for different frequencies is shown in Figure 5.3.20. The splitter operates properly for frequencies up to *100kHz*.

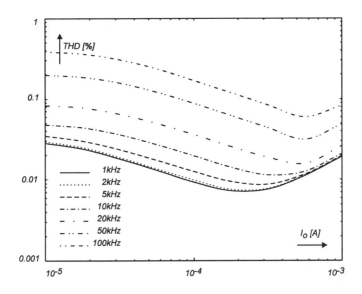

Figure 5.3.20 THD versus output amplitude for different frequencies

5.3.3 Input Stage

The implementation of the input stage is shown in Figure 5.3.21. For the input stage the same implementation is chosen as for the differential mode transconductance amplifier used in the signal splitter. The *EPMOS* and *DMOS* transistors have been cascoded in order to improve the linearity of the stage. As shown in Figure 5.3.1, the output of the transconductance amplifier is loaded with the parallel combination of resistor R_1 and capacitor C_1 that determine the dominant pole frequency of the complete amplifier. The transfer function of the input stage is given by:

$$V_d = \frac{R_1}{1+ j\omega \cdot R_1 \cdot C_1} \cdot \frac{p}{R_i} \cdot V_i$$

(5.3.39)

where p is the ratio between M_{i2} and M_{i3}. In the final realization the following values have been used: R_i=5kΩ, p=2 R_1=200kΩ, C_1=150pF, I_1=120μA and I_2=60μA.

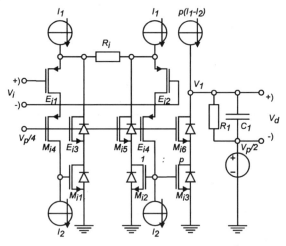

Figure 5.3.21 Input stage with dominant pole

The resulting transconductance G_1 for this stage is equal to *0.4mA/V*. The DC voltage gain equals *38dB* while the dominant pole frequency is at *5.3kHz*. The current consumption is *360μA*. The current consumption of signal splitter and input stage together is less than *4mA* which is well within the budget. With this input stage the overall transconductance G_m mounts up to *660A/V* which is within the required range.

5.3.4 Output Drivers

The output stage consists of two output drivers that copy the gate-source voltage from the replica transistors M_h and M_l to the gate-source voltage of the output transistors M_H and M_L. As explained earlier, the replica transistors M_h and M_l are enclosed in the feedback loop of the signal splitter. The output drivers are realized with two transconductance amplifiers as shown in Figure 5.3.22. In order to obtain an accurate copy V_y of the input voltage V_x the transconductance amplifiers connected in each transistor driver have to be matched as good as possible. However, the operating conditions are different for each transconductance amplifier which puts some specific requirements on the performance of the transconductance amplifiers.

The main difference between the high-side and low-side is that the low-side output transistor M_L has its source connected to ground while the high-side output transistor M_H has its source connected to the output node V_o. Therefore, the gate voltage V_H of the high-side transistor ranges from about one DMOS threshold voltage above ground to one DMOS threshold plus a few volts above the supply voltage V_P.

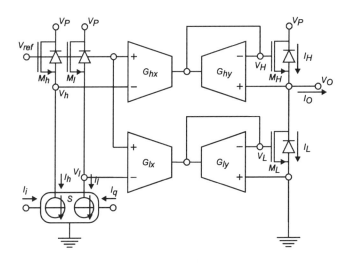

Figure 5.3.22 Output stage

The transconductance amplifiers G_{hx}, G_{lx} have their noninverting input node connected to a constant reference voltage V_{ref} while transconductance amplifier G_{ly} has its noninverting input node connected to ground. Transconductance amplifier G_{hy} on the other hand has both input nodes connected to voltages that swing up and down between, and even in excess

BCD Audio Amplifiers

of, the supply rail and ground. Consequently, the transconductance amplifiers must have almost rail-to-rail input and output windows and very high common mode rejection. Note that the nonlinearity of the transconductance is not a problem as long as both transconductance amplifiers G_x and G_y in one pair have the same transconductance. The differential input voltage can range from one DMOS threshold (~2.5V) to about 10V.

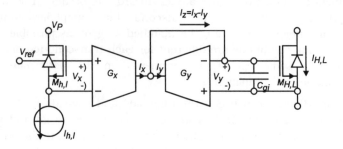

Figure 5.3.23 Output driver

The load of the voltage copier is formed by the input capacitance C_{gi} of the output transistor $M_{H,L}$ as shown in Figure 5.3.23. As explained in Section 5.3.1 the value of this capacitance ranges from about 300pF in the crossover region to almost 15nF for large output voltage excursions. For small signals, the voltage transfer of the output driver is given by:

$$\frac{V_y}{V_x} = \frac{1}{1 + j\omega \cdot \frac{C_{gi}}{G_{x,y}}}$$

(5.3.40)

The input current I_l contains many higher harmonics of the signal as a result of the signal splitting. The nonlinear voltage-to-current conversion of the replica transistor M_l creates even more higher harmonics. If the higher harmonics are attenuated by the limited bandwidth of the voltage copier this causes distortion in the output current I_L and thus in the output signal V_o.

In order to keep the distortion for audio signals low the bandwidth $G_{x,y}/2\pi C_{gi}$ of the voltage copier should be as high as possible and at least a multiple of the audio bandwidth. For example, for a bandwidth of at least 500kHz under all circumstances a minimal $G_{x,y}$ of about 50mA/V is required. The differential input voltage of the transconductance amplifiers ranges from 2.5V to 10V so the output current of the transconductance amplifiers can be as high as 500mA.

However, this poses a practical problem for the high-side driver. The output current of the high-side transconductance amplifier G_y has to be supplied by a chargepump which has a current capability of only 3mA. Therefore a different solution is required.

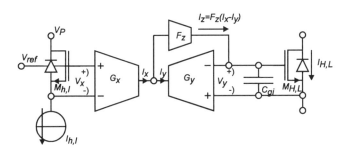

Figure 5.3.24 Improved output driver

In the output driver only the difference of the output current of both transconductance amplifiers is used to load and unload the input capacitance C_{gi}. As estimated earlier, the capacitive gate current flowing to and from the output transistors is in the order of 1mA. Consequently, the main part of the current flows from one transconductance amplifier straight into the other and is essentially wasted. This situation can be improved substantially by including a current amplifier F_z in the current copier as shown in Figure 5.3.24. The voltage transfer of this voltage copier is given by:

$$\frac{V_y}{V_x} = \frac{1}{1+j\omega \cdot \frac{C_{gi}}{F_z \cdot G_{x,y}}}$$

(5.3.41)

The bandwidth is increased by a factor F_z or, conversely, for the same bandwidth the transconductance $G_{x,y}$ can be reduced by a factor F_z. For the high-side driver only the current amplifier F_z needs to draw current from the chargepump. A current gain F_z between 10 and 100 can easily be achieved with a simple current mirror based current amplifier.

For a 500kHz bandwidth the minimal value of $F_iG_{x,y}$ is now about 50mA/V. However, for small output signals a smaller bandwidth is sufficient since the input capacitance C_{gi} is very large only for large output voltage excursions. For output voltage excursions smaller than about 20V the value of C_{gi} is smaller than 2nF. This input capacitance requires a $F_iG_{x,y}$ product of only 6mA/V to achieve a 500kHz bandwidth. The smaller value for

BCD Audio Amplifiers

$F_iG_{x,y}$ results in a bandwidth lower than *500kHz* for increasing input capacitance C_{gi}. Consequently, large output signals are distorted more than small output signals.

Implementation

An evident choice for the input stage of the transconductance amplifiers G_x and G_y is a differential pair. A suitable configuration is shown in Figure 5.3.25(a).

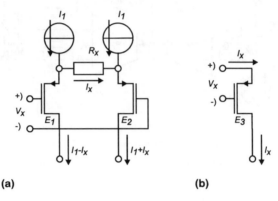

Figure 5.3.25 Transconductors (a) differential pair (b) single-ended

The choice of this particular configuration can be motivated as follows. First, the use of bipolar transistors is not possible since the base currents would distort the currents through the replica transistors. The degeneration is necessary in order to accommodate for the large input voltage range that is required. In order to have a wide common input window a Π configuration is preferred. A *p-type* differential pair is chosen to avoid the chargepump circuit in the signal path of the output drivers.

The major disadvantage of this differential pair is the sensitivity to mismatch between the tail currents I_1. The value of the currents I_1 is determined by the value of the conversion resistor R_x and the maximum input voltage $V_{x,max}$ which is about *10V*.

$$I_1 = \frac{V_{x,max}}{R_x}$$

(5.3.42)

To compensate for a current difference δI_1 in the tail currents the required input offset voltage $V_{x,offset}$ is:

$$V_{x,offset} = \delta I_1 \cdot R_x = \frac{\delta I_1}{I_1} \cdot V_{x,max}$$

(5.3.43)

A BCD Amplifier Design

So, if the mismatch between the tail currents is *1%*, the offset voltage is approximately *1%* of $V_{x,max}$ or *100mV* which is unacceptable.

An alternative is to use a single transistor as input stage as shown in Figure 5.3.25(b). In order to operate correctly the input voltage V_x has to be larger than the threshold voltage of the *(E)PMOS* which is about *1.7V*. In this application this is no restriction since the input voltage is always larger than the threshold voltage of a *DMOS* transistor which is about *2.5V*. A problem of the single transistor input stage is that the conversion current I_x is drawn from the noninverting input node. For the replica side transconductance amplifier G_x this is not a problem since this node is then connected to a reference voltage source V_{ref}. However, the input current of the output side transconductance amplifier G_y changes the voltage transfer to:

$$\frac{V_y}{V_x} = \frac{F_z}{F_z+1} \cdot \frac{1}{1+j\omega \cdot \frac{C_{gi}}{(F_z+1) \cdot G_{x,y}}}$$

(5.3.44)

As can be seen in equation (5.3.44) the DC transfer is attenuated by a factor $F_z/(F_z+1)$. This would suggest that a high value for the current gain F_z is required.

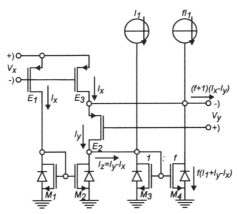

Figure 5.3.26 *Transconductance amplifier with input current correction*

However, the input current can also be compensated by using a copy of the output current of G_x. A simple implementation of an output driver is shown in Figure 5.3.26. Transistors E_1 and E_2 convert the voltages V_x and V_y into currents I_x and I_y which are then subtracted by current mirror M_1/M_2. The current amplifier F_z is implemented as a current mirror M_3/M_4 with mirror ratio f. The transistor E_3 produces a copy of I_x to compensate for the current flowing into the source of E_2. Note that the influence of mismatch between the current sources I_1 and fl_1 is divided by the loopgain F_z. The principal cause for voltage offset is now threshold mismatch between E_1 and E_2.

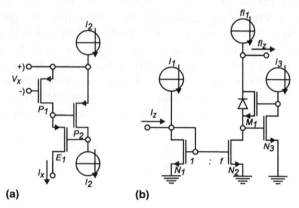

Figure 5.3.27 Active cascodes (a) input transconductor (b) current amplifier

For high common mode rejection and high accuracy, the output resistance of the conversion transistor E_1, E_2 and E_3 and the current amplifier have to be as high as possible. In order to achieve this, active cascodes are used as shown in Figure 5.3.27. Active cascodes instead of normal cascodes are used because the increase in output resistance is higher while the required saturation voltage is lower. The latter is important for the attainable voltage swing of both input and output of the output drivers.

The active cascode of the input transconductor shown in Figure 5.3.27(a) causes a bias current I_2 to be drawn through transistor P_2 from the noninverting input node. This current can be compensated for by adding an extra source I_2.

The current mirror in the current amplifier can now be realized with low voltage NMOS transistors. The NMOS transistors have a lower threshold voltage than the DMOS transistors. Consequently, the input voltage of the NMOS mirror is lower than that of a DMOS mirror. This in turn increases the allowable voltage swing of the output side transconductor G_y.

A further increase in accuracy can be obtained by using an active current mirror as shown in Figure 5.3.28. The transconductance amplifier G_z forces the drain voltage of transistor M_1 to be equal to the drain voltage of transistor M_2. This increases the accuracy of the mirror since both transistors M_1 and M_2 have the same node voltages.

A potential problem is caused by the positive feedback loop that is formed by the transconductance amplifier G_z and transistor M_2. For the stability of the circuit it is necessary that a low impedance load is connected to the mirror output node V_2. In this circuit, the load is formed by a diode connected NMOS transistor which is part of the input of the current amplifier. Due to this low impedance load the loopgain is always sufficiently low.

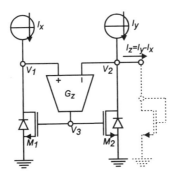

Figure 5.3.28 Active current mirror

A detailed schematic of both output drivers is shown in Figure 5.3.29. The current gain F_z of the current amplifier has been set to *24* while the transconductance of the input transconductors varies from *200µA/V* to *450µA/V*. Consequently the overall transconductance varies between *5mA/V* and *11mA/V*. In both circuits a small modification has been made to increase the output voltage window. In the low-side driver shown in Figure 5.3.29(a) a levelshift of the output voltage V_{ly} has been realized by means of DMOS transistors M_{l4} and M_{l5}. This is necessary to avoid saturation of the transconductor composed of P_{l5}, P_{l6} and E_{l3}. The DMOS transistors cause a voltage shift of about *3V*.

In the high-side driver shown in Figure 5.3.29(b) the compensation for the input current of G_y is not added directly at the input but at the internal node V_{h1} of the current amplifier. This is necessary in order to allow the common level of the output voltage V_{hy} to exceed the common level of the input voltage V_{hx}. Note that only the current source I_h is supplied by the chargepump.

The bias current of the low-side driver is *3.7mA* while the high-side driver draws *0.7mA* from the power supply V_P and *2.9mA* from the chargepump. If it is assumed that the chargepump consumes about *2.5* times the current it supplies the combined bias current is approximately *11.6mA*.

The bias current of the input stage, signal splitter and output drivers adds up to about *15.6mA* which is close to the specified value of *15mA*.

BCD Audio Amplifiers

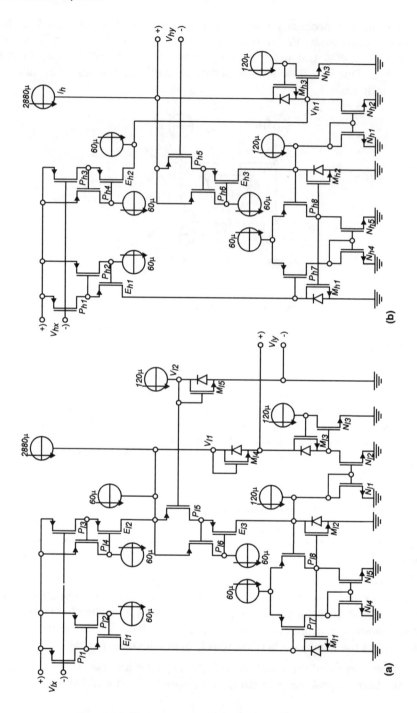

Figure 5.3.29 Output drivers (a) low-side (b) high-side

Simulation Results

The currents I_H and I_L through the high-side and low-side output transistors for a *100W/4Ω, 1kHz* sinewave output signal are shown in Figure 5.3.30.

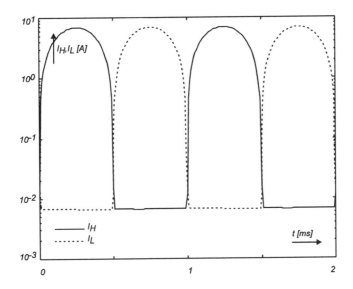

Figure 5.3.30 High-side and low-side currents I_H and I_L

As can be seen the residual value of both output currents are limited to about *6mA* so the output transistors are conducting all the time. In Figure 5.3.31 the distortion of the output current $I_o = I_H - I_L$ for the same output signal is shown. For large signals such as this the large signal distortion dominates while crossover distortion is hardly distinguishable. Note that the distortion of the signal is almost symmetrical which means that even order harmonic distortion is relatively small.

In Figure 5.3.32 the voltage transfer V_y/V_x of the low-side driver as a function of frequency is shown for different values of the output voltage V_o.

The increase of the input capacitance for large output voltage excursions results in a significant reduction in bandwidth of the driver. For output voltage excursions up to about *15V* the bandwidth of the driver is over *1MHz*. For larger excursions, the increasing input capacitance causes the bandwidth to drop to about *100kHz* which is only *5* times the audio bandwidth. Therefore, it can be expected that the distortion increases for higher output power levels.

BCD Audio Amplifiers

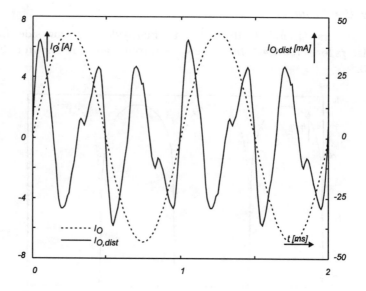

Figure 5.3.31 Open loop distortion of the output current $I_O=I_H-I_L$

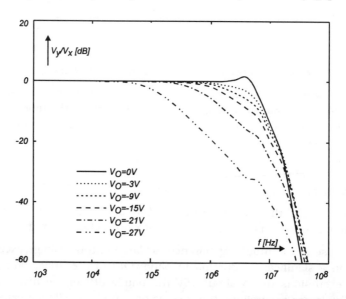

Figure 5.3.32 Bandwidth of low-side driver for different output voltages V_o

The total harmonic distortion as a function of output power for different frequencies is shown in Figure 5.3.33. As can be seen in Figure 5.3.33 the THD increases moderately for output power up to about 50W and starts to increase rapidly for higher output power.

A BCD Amplifier Design

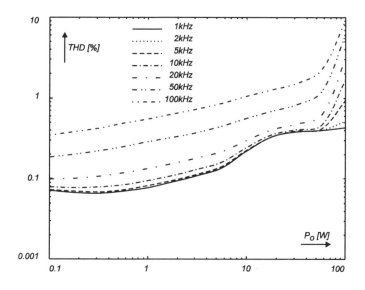

Figure 5.3.33 THD versus output power for different frequencies

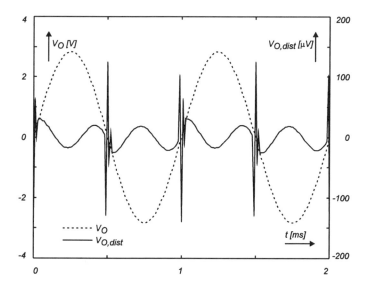

Figure 5.3.34 Distortion of a 1kHz 1W/4Ω sinewave

For a 4Ω load the voltage amplitude of a 50W sinewave is 20V. As can be seen in Figure 5.3.3 the input capacitance of the output transistors starts

195

BCD Audio Amplifiers

to increase rapidly at this output voltage due to the accumulation effect. Consequently, the input capacitance appears to be the dominant source of distortion in the output stage.

5.3.5 Complete Amplifier

In this section simulation results are presented of the complete amplifier. The amplifier is composed of the input stage, signal splitter and output stage presented in the previous sections plus a resistive feedback network. The closed loop gain of the amplifier is set at *30dB*. In Figure 5.3.34, Figure 5.3.35 and Figure 5.3.36 the output signal V_O and distortion $V_{O,dist}$ are shown for a 1kHz sinewave with output power of *1W/4Ω*, *10W/4Ω* and *100W/4Ω* respectively.

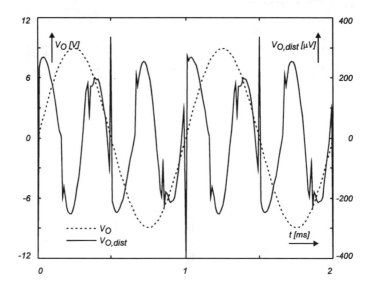

Figure 5.3.35 Distortion of a 1kHz 10W/4Ω sinewave

As can be seen in Figure 5.3.34 the distortion is dominated by crossover distortion for low output power levels. For higher frequencies the magnitude of the crossover peaks increases proportionally while the large signal distortion remains almost constant. At *1kHz* the difference between signal amplitude and crossover peak value is about *86dB*. For moderate power levels the large signal distortion becomes significant as can be seen in Figure 5.3.35. In this case the magnitude of the crossover peaks is about equal to the large signal distortion. The small peaks in the distortion $V_{O,dist}$ outside the crossover region are caused by a discontinuity in the output transistor model. Similar to the low power case the magnitude of the crossover peaks increases almost proportionally with frequency.

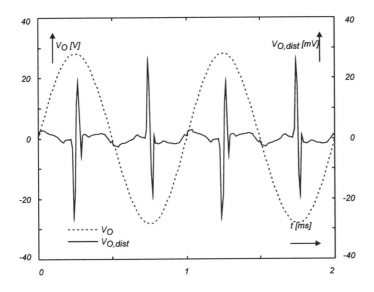

Figure 5.3.36 Distortion of a 1kHz 100W/4Ω sinewave

For high output power the large signal distortion dominates, as can be seen in Figure 5.3.36. The large peaks in the distortion $V_{O,dist}$ are the result of the output transistors going out of saturation.

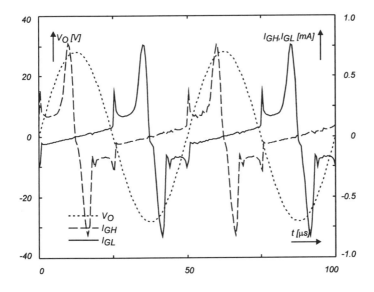

Figure 5.3.37 Gate currents of the output transistors for a 20kHz 100W/4Ω sinewave

In Figure 5.3.37 the gate currents of the output transistors are shown for a 20kHz, 100W/4Ω sinewave output signal. Note that these currents are almost symmetrical and also identical for the low-side and high-side transistors. As can be seen in Figure 5.3.37 peak currents occur during crossover and in the peaks of the output signal. In the crossover region peak currents occur in the gates of both output transistors due to the short moment of class A operation during crossover. The large gate currents in the peaks of the output signal are caused by the strong increase of the input capacitance of the output transistors.

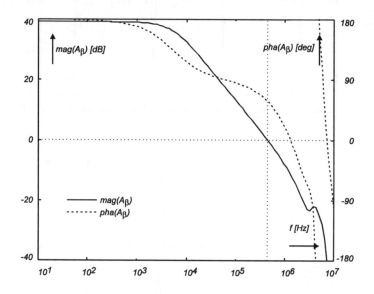

Figure 5.3.38 Bodeplot of the loopgain A_β

In Figure 5.3.38 and Figure 5.3.39 bodeplots of the loopgain A_β and output resistance R_o of the complete amplifier are shown. Both plots are obtained by performing an AC analysis in the quiescent point of the amplifier. For different operating points the loopgain A_β and output resistance R_o are also different. However, the dominant characteristics remain unchanged. As can be seen in Figure 5.3.38 the dominant pole frequency of the loopgain A_β is located at about *5kHz* while the unity-gain frequency is at about *450kHz*. The phasemargin and gainmargin are *58°* and *10dB* respectively. Note that the loopgain depends on the load resistor R_L which is 4Ω in this case. For an 8Ω load the gain increases by *6dB* resulting in a phasemargin and gainmargin of about *31°* and *4dB* respectively. For these small margins it may be necessary to reduce the dominant pole frequency or apply Miller compensation. In this form the amplifier is not stable if no load is connected to the output.

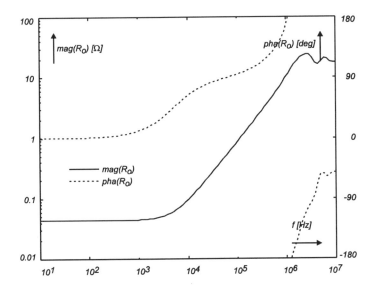

Figure 5.3.39 Bodeplot of the output resistance R_o

The output resistance R_o is $44m\Omega$ for low frequencies and starts to increase at the dominant pole frequency. At *20kHz* the magnitude of the output resistance is *$176m\Omega$*.

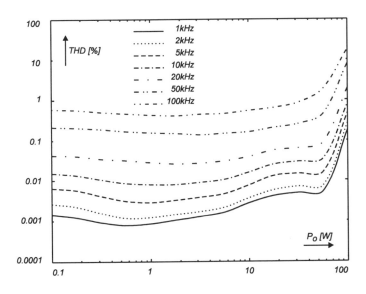

Figure 5.3.40 THD versus output power for different frequencies

Finally, the total harmonic distortion of the complete amplifier as a function of output power for different frequencies is shown in Figure 5.3.40. The *THD* curves all show a sharp increase for output power higher than $50W/4\Omega$. Therefore, it can be concluded that the dominant source of distortion is the increase of the input capacitance of the output transistors for large output signal excursions. This effect is inherent to the *VDMOS* transistor and can be expected to be the dominant source of distortion in other amplifier topologies as well.

Clearly, the validity of these simulation results still have to be demonstrated with measurements. Unfortunately, no relevant measurement data was available at the time of print. However, the simulations are expected to give a reasonable indication to what is possible in an actual realization.

5.4 Conclusion

Distortion in audio amplifiers is dominated by the output stage and is caused by large signal nonlinearity in the characteristics of the used components and the nonlinear nature of class *AB* push-pull operation.

A comparison of various output stage topologies with *n-DMOS* output transistors has been presented. A distinction is made in common drain stages and common source stages. Common drain stages have a low open loop output resistance and are used as a voltage buffer. Common source stages have a high open loop output resistance and are used as transconductance amplifiers. The closed loop output resistance is forced low by using voltage feedback.

One output stage topology called an *improved current amplifier* has been selected for further development. It is expected that the stability of this stage is good since, as opposed to other topologies, it does not include the output transistors in a local feedback loop to control the quiescent and residual currents. The design concentrates on two parts: a *signal splitter* and two *output drivers*.

An active signal splitter design based on translinear circuit principles and active feedback has been presented. The output currents of the splitter approximate a harmonic mean law which results in smooth crossover and favorable residual current levels. The distortion of this active splitter has been shown to be substantially lower than that of a conventional passive signal splitter. Some techniques are presented to reduce large signal distortion and crossover distortion in the active signal splitter while maintaining moderate loopgain in the feedback loop. Simulations of the signal splitter show low distortion levels and an almost symmetrical distortion behavior which is the result of the perfect symmetry in the signal splitter circuit.

Two almost identical output driver designs are presented that use two transconductance amplifiers to copy the gate-source voltage from the replica transistors to the gate-source voltage of the output transistors. The transconductance amplifiers are based on single-ended VI conversion with a single MOS transistor. Various refinements in the output driver circuits have been presented to improve the accuracy and common mode rejection. The almost identical designs of the low-side and high-side output driver result in symmetrical distortion.

Simulations of the complete amplifier have shown that total harmonic distortion is well below *0.1%* for low and moderate output power levels for signals within the audio band. For high output power levels the large input capacitance of the output transistors cause a fast increase in distortion. Although no measurement data was available, the simulations give an indication of what is possible in an actual realization.

5.5 References

[1] Bult, K. *Analog CMOS Square-Law Circuits*, Ph.D. Thesis, University of Twente, 1988
[2] Self, D., "Distortion in Power Amplifiers 1: The Sources of Distortion", *Electronics World & Wireless World*, Vol.99, pp.630-634, Aug. 1993.
[3] Self, D., "Distortion in Power Amplifiers 2: The Input Stage", *Electronics World & Wireless World*, Vol.99, pp.730-736, Sep. 1993.
[4] Self, D., "Distortion in Power Amplifiers 3: The Voltage-Amplifier Stage", *Electronics World & Wireless World*, Vol.99, pp.818-824, Oct. 1993.
[5] Self, D., "Distortion in Power Amplifiers 4: The Power-Amplifier Stages", *Electronics World & Wireless World*, Vol.99, pp.929-934, Nov. 1993.
[6] Self, D., "Distortion in Power Amplifiers 5: Output Stages", *Electronics World & Wireless World*, Vol.99, pp.1009-1014, Dec. 1993.
[7] Self, D., "Distortion in Power Amplifiers 6: The Remaining Distortions", *Electronics World & Wireless World*, Vol.100, pp.41-45, Jan. 1994.
[8] Self, D., "Distortion in Power Amplifiers 7: Frequency Compensation and Real Designs", *Electronics World & Wireless World*, Vol.100, pp.137-142, Feb. 1994.
[9] Self, D., "Distortion in Power Amplifiers 8: Class A Amplifiers", *Electronics World & Wireless World*, Vol.100, pp.225-231, Mar. 1994.
[10] Cherry, E.M., "Ironing out Distortion", *Electronics World & Wireless World*, Vol.101, pp.14-20, Jan. 1995
[11] Seevinck, E., W. de Jager, P. Buitendijk, "A Low-Distortion Output Stage with Improved Stability for Monolithic Power Amplifiers", *IEEE Journal of Solid-State Circuits*, Vol.23, No.3, pp.794-801, Jun. 1988
[12] Cherry, E.M., Nested Differentiating Feedback Loops in Simple Audio Power Amplifiers", *Journal of the Audio Engineering Society*, Vol.30, No.5, pp.295-305, May. 1982.
[13] Brasca, G., E. Botti, "A 100V/100W Monolithic Power Audio Amplifier in Mixed Bipolar-MOS Technology", *IEEE Transactions on Consumer Electronics*, Vol.38, No.3, pp.217-222, Aug. 1992
[14] Wiegerink, R.J., *Analysis and Synthesis of MOS Translinear Circuits*, Ph.D. Thesis, University of Twente, 1992
[15] Lish, A., "A Class A/B Floating Buffer BiCMOS Power Op-Amp", *IEEE Journal of Solid-State Circuits*, Vol.30, No.6, pp.670-676, Jun. 1995

[16] Verboon, A.W., *A Current Driven BiCMOS-DMOS Audio Amplifier Output Stage*, M.Sc. Thesis, Universitry of Twente, 1996
[17] Gray, P.R., R.G. Meyer, *Analysis and Design of Analog Integrated Circuits*, 2nd Edition, John Wiley and Sons, New York, 1984
[18] Linsley Hood, J., "Solid-State Audio Power I", *Electronics World & Wireless World*, Vol.95, pp.1024-1048, Nov. 1989
[19] Linsley Hood, J., "Solid-State Audio Power II", *Electronics World & Wireless World*, Vol.95, pp.1164-1168, Dec. 1989
[20] Linsley Hood, J., "Solid-State Audio Power III", *Electronics World & Wireless World*, Vol.96, pp.16-21, Jan. 1990

6

Conclusion

In this book the design of a fully integrated *100W* audio power amplifier in a *BCD* technology is presented.

An attempt has been made to develop a design strategy that can be applied to *BCD* technologies in general. However, since the designs presented in this book are all realized in one specific *BCD* technology, many design choices are based on the characteristics of this technology.

Key elements in the amplifier design are a fully integrated chargepump circuit and a common source output stage. The overall design goal has been to achieve high open loop linearity.

6.1 Conclusions

In *Chapter 1* a number of amplifier classes has been presented and compared with respect to power dissipation in the output transistors of the amplifier. Single-ended class *AB* amplifiers form the basis of many other amplifier classes and are a suitable vehicle for research. A list of specifications has been given that an integrated audio amplifier has to satisfy. In order to achieve rail-to-rail output capability, it is necessary to have a voltage higher than the supply voltage available to drive the gate of the high-side output transistor.

Conclusion

In *Chapter 2* a qualitative overview based on an extensive literature study has been presented of the various device structures, the device physics and the technological aspects of DMOS transistors. A comparison between DMOS transistors and bipolar transistors has yielded that the most important advantage of DMOS transistors compared to bipolar transistors is the excellent thermal stability and the immunity to second breakdown.

In *Chapter 3* it has been shown that an integer multiple of the supply voltage can be generated without the use of inductors by using switched capacitor techniques. Three well-known voltage multiplication techniques have been presented and compared on their suitability for integration. Based on this comparison a specific voltage multiplier circuit called a *chargepump* has been selected for application in *DMOS* output stages. At the output of a chargepump a considerable ripple voltage occurs which is proportional to the load current of the chargepump. The magnitude of this voltage ripple can be reduced substantially by driving the chargepump with two switched current sources operating in antiphase instead of a squarewave clock voltage. This new mode of operation is called *double phase current driven* mode. An additional advantage of this new driving technique is that the current drawn from the supply voltage is almost constant in contrast to the high peak currents that occur in conventional chargepump circuits.

A fully integrated chargepump design has been presented in which the reduction of the output voltage ripple is demonstrated. It has been shown that it is possible to realize a chargepump capable of sourcing a few milliAmperes of current at an output voltage *10V* higher than the supply voltage within a reasonable chip area.

In an appendix it has been shown that switched capacitor techniques can also be used to realize division of the supply voltage. The output voltage ripple of such a voltage divider can be reduced using a similar driving mode as used for chargepumps. Ripple reduction in voltage multipliers and voltage dividers can also be applied for power *DC-DC* conversion if external capacitors are used.

In *Chapter 4* the development of a new extended chargepump model has been presented which is based on analysis of the charge balance between two adjacent capacitors in a chargepump circuit. This model describes the steady-state and the transient behavior of chargepump circuits as well as the power consumption and conversion efficiency. A number of parasitic effects can be included into the model by simple modifications in the charge balance analysis. The validity of the model has been demonstrated with circuit simulations. Further, it has been demonstrated that a number of previously published chargepump models can be derived from the new extended model.

Conclusions

In *Chapter 5* the design of an integrated *100W* audio power amplifier realized in a *BCD* technology has been presented. Several different amplifier topologies have been discussed and compared with respect to *distortion, stability* and *quiescent current control*. A distinction is made between quasi-complementary *common source* and quasi-complementary *common drain* stages. The common drain stages have a low open loop output resistance but all require local feedback loops around the output transistors which is a potential source of instability. In the common source stages, on the other hand, the open loop output resistance is high but no local feedback is required around the output transistors. Therefore, it is expected that the stability of these stages is good. The design of an amplifier with a common source output stage concentrates on two parts: a class *AB signal splitter* and two *output drivers*.

An new *active signal splitter* design has been presented that is based on translinear circuit principles and active feedback. The splitter divides the signal current in a *high-side* and a *low-side* current that are related by a *harmonic mean law* which results in a smooth crossover and a favorable residual current level that avoids turn-off of the output transistors. In the signal splitter a combination of feedback and compensation is applied to achieve a very high ratio (*1:1000*) between the signal amplitude and quiescent current level. A comparison with a conventional signal splitter shows a substantial reduction in both *large signal distortion* and *crossover distortion* of the active signal splitter. Simulations of the *total harmonic distortion* (*THD*) show that the distortion caused by the splitter can be considered negligible compared to the distortion caused by the output current amplifier.

The output currents of the signal splitter are amplified separately and summed at the output node. A new *output driver design* has been presented that copies the gate-source voltage of a small *replica* transistor to the gate-source voltage of an output transistor. With some minor modifications, the same output driver design can be applied to realize current amplification for both the high-side and the low-side. The high-side output driver is constructed in such a way that it draws a constant current from the chargepump circuit.

Quiescent current consumption of the complete amplifier is approximately *25mA*. Simulations of the complete amplifier show that total harmonic distortion is well below *0.1%* for low (*1W/4Ω*) and moderate (*10W/4Ω*) output power for signals within the audio band. For high (*100W/4Ω*) output power an increase in distortion has been found. This is caused by a strong increase in the input capacitance of the output transistors which results from an *accumulation* phenomenon that is particular to Vertical *DMOS* transistors.

6.2 Recommendations

In this book only one amplifier topology has been designed in detail which was selected because it is expected to exhibit good stability and low distortion. However, in order to make a fair comparison it would be interesting to design one or more of the other topologies as well.

The presented amplifier has a single-ended (*SE*) configuration. A bridge-tied-load (*BTL*) configuration has a four times higher output power capability while at the same time it does not require a decoupled buffered half supply voltage or coupling capacitor for the load. The extension of a *SE* configuration to a *BTL* configuration involves the design of a common mode feedback loop.

The switched capacitor technique used in the chargepump circuit can be used to realize both *step-up* and *step-down DC-DC* conversion. Step-up and step-down *DC-DC* converters for higher output power can be realized by using external capacitors. In these converters the same ripple reduction technique presented in this book can be applied. The *DC* voltages that are generated by the converters can be used as extra voltages to supply class *G* amplifiers.

The increase in input capacitance of the output devices is caused by an accumulation phenomenon that is particular to Vertical *DMOS* transistors. However, as has been mentioned in *Chapter 2*, this capacitance can be reduced by selectively removing the gate or using a thick oxide at the drift region overlap. Further, it may be advantageous to use Lateral *DMOS* transistors instead of Vertical *DMOS* transistors.

Index

A

accumulation region · 25
active cascode · 190
active signal splitter · 173
active stage control · 82
amplifier class
    class A · 4; 5
    class AB · 7
    class B · 6; 7
    class D · 9
    class G · 7
amplitude control · 81
avalanche breakdown · 20; 54
average dissipation · 3; 4
average output power · 3

B

backgate · 21
BCD technology · 12; 13; 145
bidirectional switch · 38
bipolar transistor · 12; 50
body-effect · 99; 102; 128
boost conversion · 63
bootstrap · 64
bootstrap capacitor · 14
breakdown · 25; 30; 34
bridge-tied-load (BTL) · 3
BTL · *See* bridge-tied-load
bulk · 18

C

channel · 19
channel length · 21
channel length modulation · 19; 38

chargepump · 14; 73; 97; 100; 101; 165; 187; 191
    current driven · 78; 95
    double phase · 76; 94
    double phase current driven · 78
charge balance · 101
charge balance equation · 104
closed loop gain · 10; 196
closed loop voltage gain · 142; 163
Cockcroft-Walton voltage multiplier · 70
common drain configuration · 146
common source configuration · 146
composite-PNP · 147
conversion efficiency · 132; 133
crossover distortion · 6; 7; 11; 12; 139; 140
current amplifier · 154; 156; 159; 162; 163; 169
current gain · 163; 168
current leakage · 112
current limit · 52

D

deadband · 141
depletion capacitance · 123
Dickson voltage multiplier · 72
dielectric isolation · 43
dissipation · 1; 3; 4; 10; 138; 164
    class A · 5
    class AB · 7
    class B · 6
    class D · 9
    class G · 7; 8
dissipation limit · 52
distortion · 4; 11; 138; 162; 163; 182; 193; 196; 200
distortion spectrum · 142
DMOS transistor · 12; 13; 21

doping profile · 36
double-diffusion · 21; 36
drift region · 20
dynamic crossover distortion · 141; 172
dynamic large signal distortion · 140

E

EEPROM · 66
efficiency · 1; 4; 5; 63; 93; 100
epitaxial layer · 23
equivalent circuit
   DMOS transistor · 26; 31; 35
   voltage multiplier · 66
equivalent output resistance · 66; 68; 71; 73; 99; 118
extended drain · 20
extended drain PMOS · 48

F

feedback factor · 142
field plate · 26; 44
field ring · 26; 44
floating buffer stage · 151
floating field ring · 44
floating signal splitter · 156
feedback factor · 142

G

gain factor · 37
gate-drain capacitance · 40; 166; 167
gate-source capacitance · 39; 166; 167
geometric mean · 147
gm-doubling · 141
ground state · 102
grounded signal splitter · 156

H

harmonic distortion · 7; 139
harmonic mean · 153; 169; 173
hot-carrier injection · 20
hot spot · 53
hybrid devices · 55

I

input capacitance · 13; 18; 39; 167; 168; 200
input resistance · 18
input stage · 10; 143; 168; 184
instantaneous dissipation · 2
instantaneous output power · 2
insulated base transistor (IBT) · 56
insulated gate transistor (IGT) · 56
intelligent power device (IPD) · 22
inversion · 19; 36
isolation · 41

J

JFET · 25
junction capacitance · 123
junction curvature · 44
junction isolation · 42
junction temperature · 53
junction termination · 44
junction termination extension · 44

L

large signal distortion · 139
lateral DMOS (LDMOS) · 23
lightly doped drain · 20
linear region · 19
linearity · 12
loop gain · 169; 198

M

Marx voltage multiplier · 68
maximum output power · 3
Miller-effect · 167; 168
minimum selector · 153; 161; 162; 163
minority carrier charge storage · 18
mismatch · 37; 79; 140; 179; 188

N

neck region · 29

Index

O

offset · 158; 179; 188
on-resistance · 13; 18; 25; 30; 34; 164
open-base breakdown · 53
output driver · 10; 166; 187
output power · 3; 5; 10; 134; 164
output resistance · 11; 138; 142; 164; 198
output voltage clipping · 81
oxide breakdown · 20

P

parallel capacitance · 98; 122
parasitic bipolar transistor · 40; 54
passive signal splitter · 170
pinch-off · 19
power integrated circuit (PIC) · 22
power consumption · 132
process flow · 45
proximity effect · 30
pulse-width-modulation (PWM) · 9; 64
pump capacitor · 101
punchthrough · 20
push-pull · 2
PWM · *See* pulse-width-modulation

Q

quasi-complementary stage · 147; 149; 151
quasi-saturation · 30
quiescent current · 2; 5; 6; 10; 138; 162; 164; 170; 181
quiescent current control · 148; 162

R

reactive harmonic distortion · 140
reduced surface field (RESURF) · 24
reliability · 18; 52
replica transistor · 140; 152; 165; 168; 185
residual current · 7; 138; 162

S

safe operating area (SOA) · 12; 18

SE · *See* single-ended
second breakdown · 12; 18; 52; 54
self isolation · 41
self-heating · 43
series resistance · 99; 114
signal splitter · 154; 162; 168; 169
single-ended · 2
slew-rate · 11; 165
smart power technology · 22
SOA · *See* safe operating area
source-drain diode · 38
source follower · 146
square law characteristic · 12; 37
stability · 12; 145; 162; 163; 169
start-up behavior · 100; 113; 116; 120; 125; 130
state transition equation · 105
state variable · 102; 105
static crossover distortion · 141
static large signal distortion · 139
steady-state output voltage · 68; 70; 72; 74; 75; 76; 77; 78; 93; 106; 107; 113; 117; 121; 125; 130
storage capacitor · 65; 101
storage capacitance · 123
stray capacitance · 119
step-up conversion · 63
step-down conversion · 91
subthreshold region · 36
switching distortion · 141
switching speed · 12; 18
switching voltage regulator · 64

T

THD · *See* total harmonic distortion
thermal lateral instability · 53
thermal stability · 12
thermal resistance · 43; 164
threshold voltage · 13; 18; 36; 127
total harmonic distortion (THD) · 11; 183; 194
totempole configuration · 13; 137
transconductance · 12; 18; 144; 164; 166; 167; 168
transconductance amplifier · 143; 154; 161; 163
transconductance stage · 146
transfer characteristic · 12; 37
transient behavior · 100

209

translinear loop · 147; 150; 152; 170

U

U-groove MOS (UMOS) · 34

V

velocity saturation · 30
vertical DMOS (VDMOS) · 27
V-groove · 33
V-groove MOS (VMOS) · 32
virtual output power · 5
voltage amplifier · 142
voltage buffer · 146; 162
voltage copier · 150; 186
voltage divider · 91
voltage multiplier · 65
voltage ripple ·75; 77; 79; 92; 98; 109; 114; 118; 122; 126; 131;